Energy Harvesting Solutions for Implantable Medical Devices

Other related titles:

You may also like

- PBCS078 | Iqbal, Aïssa, and Nauman | Energy Harvesting for Wireless Sensing and Flexible Electronics through Hybrid Technologies | 2023
- PBHE011 | Velez, and Miyandoab | Wearable Technologies and Wireless Body Sensor Networks for Healthcare | 2019
- PBHE010 | Goleva, Ganchev, Dobre, Garcia, and Valderrama | Enhanced Living Environments: From models to technologies | 2017
- SBEW555 | Whittow | Bioelectromagnetics in Healthcare: Advanced sensing and communication applications | 2022

We also publish a wide range of books on the following topics:
Computing and Networks
Control, Robotics and Sensors
Electrical Regulations
Electromagnetics and Radar
Energy Engineering
Healthcare Technologies
History and Management of Technology
IET Codes and Guidance
Materials, Circuits and Devices
Model Forms
Nanomaterials and Nanotechnologies
Optics, Photonics and Lasers
Production, Design and Manufacturing
Security
Telecommunications
Transportation

All books are available in print via https://shop.theiet.org or as eBooks via our Digital Library https://digital-library.theiet.org.

IET HEALTHCARE TECHNOLOGIES SERIES 052

Energy Harvesting Solutions for Implantable Medical Devices

Design, integration, and application of self-powered biomedical implants

Jinwei Zhao and Hadi Heidari

Institution of Engineering and Technology

About the IET

This book is published by the Institution of Engineering and Technology (The IET).

We inspire, inform and influence the global engineering community to engineer a better world. As a diverse home across engineering and technology, we share knowledge that helps make better sense of the world, to accelerate innovation and solve the global challenges that matter.

The IET is a not-for-profit organisation. The surplus we make from our books is used to support activities and products for the engineering community and promote the positive role of science, engineering and technology in the world. This includes education resources and outreach, scholarships and awards, events and courses, publications, professional development and mentoring, and advocacy to governments.

To discover more about the IET please visit https://www.theiet.org/.

About IET books

The IET publishes books across many engineering and technology disciplines. Our authors and editors offer fresh perspectives from universities and industry. Within our subject areas, we have several book series steered by editorial boards made up of leading subject experts.

We peer review each book at the proposal stage to ensure the quality and relevance of our publications.

Get involved

If you are interested in becoming an author, editor, series advisor, or peer reviewer please visit https://www.theiet.org/publishing/publishing-with-iet-books/ or contact author_support@theiet.org.

Discovering our electronic content

All of our books are available online via the IET's Digital Library. Our Digital Library is the home of technical documents, eBooks, conference publications, real-life case studies and journal articles. To find out more, please visit https://digital-library.theiet.org.

In collaboration with the United Nations and the International Publishers Association, the IET is a Signatory member of the SDG Publishers Compact. The Compact aims to accelerate progress to achieve the Sustainable Development Goals (SDGs) by 2030. Signatories aspire to develop sustainable practices and act as champions of the SDGs during the Decade of Action (2020-30), publishing books and journals that will help inform, develop, and inspire action in that direction.

In line with our sustainable goals, our UK printing partner has FSC accreditation, which is reducing our environmental impact to the planet. We use a print-on-demand model to further reduce our carbon footprint.

British Library Cataloguing in Publication Data

A catalogue record for this product is available from the British Library

ISBN 978-1-83953-685-4 (hardback)
ISBN 978-1-83953-686-1 (PDF)

Typeset in India by MPS Limited

Cover image: Christoph Burgstedt/Science Photo Library via Getty Images

Contents

About the authors

Jinwei Zhao is a senior electronics engineer at Imperial Brands PLC, focusing on next-generation heating tobacco devices and low-temperature vaping systems. He holds a PhD in Electrical and Electronics Engineering from the University of Glasgow. His previous roles include senior bioelectronics engineer at QV Bioelectronics and Research assistant at the State Key Laboratory of Analog and Mixed-Signal VLSI (University of Macau). He has published extensively in IEEE and Wiley journals and received the IEEE PRIME Silver Leaf Award and Wiley's Most Downloaded Paper Award.

Hadi Heidari is a professor of nanoelectronics in the James Watt School of Engineering, University of Glasgow, UK. He is a senior member of IEEE and a member of the RSE Young Academy of Scotland. He has published over 300 peer-reviewed publications, patents, book chapters and presentations and has been on the technical committee and organising committees of numerous international conferences. He has been the recipient of several awards, including the 2020 IET's JA Lodge Award.

Chapter 1
Background of implantable bioelectronics device and applications

An implantable medical device (IMD) is characterized as an apparatus that is either partially or entirely integrated into the human body through surgical or medical procedures, remaining in situ post-implantation for an extended period. These devices or applications undoubtedly assist patients in restoring bodily functions compromised by conditions such as hearing loss, heart failure, and Parkinson's disease, thereby enhancing their quality of life and prolonging lifespan. The field of implantable biomedical devices has witnessed significant advancements within the domains of microelectronics, biotechnology, and biomaterials. Such devices are capable of executing intricate functions, including vital real-time monitoring, endovascular repair, and neural stimulation, which have already alleviated chronic ailments for millions of patients. This chapter elucidates the current utilization of implantable devices in contemporary hospital settings, examines the development of these emergent applications, delineates the methods for device validation, and addresses the quality and risk regulation control. Furthermore, we discuss the future prospects of implantable devices.

1.1 Situation of implantable device in the modern clinic

According to the definition provided by the United States Food and Drug Administration (FDA), a medical device constitutes a category of healthcare products capable of delivering health treatments or services unattainable via chemical means or metabolic processes [7]. Within the FDA's classification system, IMDs are identified as Class III due to their high-risk nature. In contrast to Class I and II devices, which are generally perceived to have a lower safety threshold for market implementation, Class III devices necessitate stringent clinical and technical testing. Consequently, the design and manufacturing of Class III devices present greater challenges and incur substantial validation costs prior to implantation in the human body. Jiang *et al.* reports that 8%–10% of the American population, as well as 5%–6% of the industrialized countries' populations, utilized IMDs for healthcare and treatment by the year 2009 [45,48]. The market for IMDs continues to expand, with new features being continually updated in response to the growing demand for safer, higher-quality life care from patients, in conjunction with the clinical requirements of device providers [64].

The aging population and the rise in chronic disease incidences contribute to the increasing number of individuals who could benefit from IMDs [59]. In a report by K.D. Lind in 2017, it was illustrated that 370,000 cardiac pacemakers were received by American patients, and it was estimated that approximately 7.2 million American patients relied on joint implantable devices [59]. In the United States, expenditures on medical devices exceeded $170 billion and accounted for about six percent of total national health expenditures of $2.9 trillion in 2013.

1.2　Implantable applications in the modern therapy and treatment

Dynamic power-performance management encompasses energy harvesting, storage, and voltage conversion. These elements work together to enhance the lifespan of implantable devices, where energy harvesting and storage are crucial. Voltage conversion optimizes the voltage and current needs of the device loads [16].

Energy-efficient signaling involves low-power analog and digital signal processing. The analog processing part primarily cuts down the power needs of analog-to-digital conversion by adjusting for dynamic range and device inconsistencies. To counteract power loss from digital circuit noise margins and distortions, the digital processing provides adaptable energy-efficient architectures. These strategies aim to prolong device longevity while minimizing power loss during load and signal processing [16].

Reducing power usage and losses is key to extending the lifespan of implantable biomedical devices. A study illustrated that lowering an implantable device's power usage from 10 mW to 8 μW can extend its lifespan from 3 days to 10 years [16,100]. Employing low-power components boosts device functionality, though more functional blocks increase power demands. Consequently, the selection of power harvesting technology is application-specific, requiring thorough analysis of implantable applications' specifications and power needs.

Health applications predominantly use devices like cardiac pacemakers, cochlear implants, drug pumps, retinal, and neural stimulators. Invented in 1958, pacemakers have aided over 200,000 patients with cardiac issues [21,78]. Pacemakers typically last 7–10 years, requiring battery recharging instead of costly surgical replacements (cost: $6000–$12,000) [16,21,83]. Their power usage generally falls in the microwatt range with minimal processing and slow analog-to-digital speeds. Some pacemakers, however, demand high-energy pulses for defibrillation [21].

In 1961, W. House introduced cochlear implants, aiding two patients with hearing impairments. Designed for restoring hearing via electric stimulation, over 200,000 people have benefitted from implantable hearing aids, about 250,000 with pacemakers in the USA [6,89]. The evolution of power management technologies significantly contributes to the hearing aid market [105]. Cochlear implants consume between 100 and 2000 μW. In comparison, retinal and neural stimulators, as emerging technologies, vary in power consumption from micro- to milliwatts depending on smart array setups. Stenevi first developed neurological implants in mammals in 1976 [9].

Similarly, retinal implants were developed to help regain basic visual perception for those with vision impairment. The initial designs and implementations were carried out by Brindley and Lewin in 1968 [99]. Much like neural stimulators, the electrical power needs of retinal stimulators are linked to the layout of the electrode array. Originating in the 1960s, the drug pump system was created as an alternative to oral drug delivery in certain treatments [10]. Unlike earlier approaches, this drug delivery system can be placed in various bodily regions, such as intraocular, intrauterine, and vaginal sites. Once the drug is depleted, the system must be surgically extracted, posing a challenge for it to be a long-term self-powered device [51]. In the sections that follow, we outline the distinct energy harvesting approaches employed in different implantable applications, discussing the operational principles of each technology and evaluating how they satisfy the power needs of implantable electronic devices.

1.3   Life cycle evidence in implantable devices

R. Tarricone, in 2020, elaborates on the "Evidence Council on Pathways to Clinical Evidence Generation for High-Risk Implantable Medical Devices," initiated and directed by the Center for Research on Health and Social Care Management (CERGAS) [87]. The outlined methodology includes stages of pre-clinical and pre-market evidence collection, clinical pre-market evidence compilation, the diffusion phase, and post-market evidence activities, concluding with the obsolescence and replacement stage of post-market clinical evidence gathering. Figure 1.1 [87] illustrates the pre-clinical research phase, which generally involves biocompatibility and toxicity assessments crucial for device design optimization, prototype development, and engineering production [87]. This phase is pivotal for defining the design process and evaluating related risks [1]. Current evidence highlights the role of computational modeling in the pre-clinical phase, offering simulations of scenarios that inform, the design and prediction of high-risk IMDs for specific patients in distinct clinical

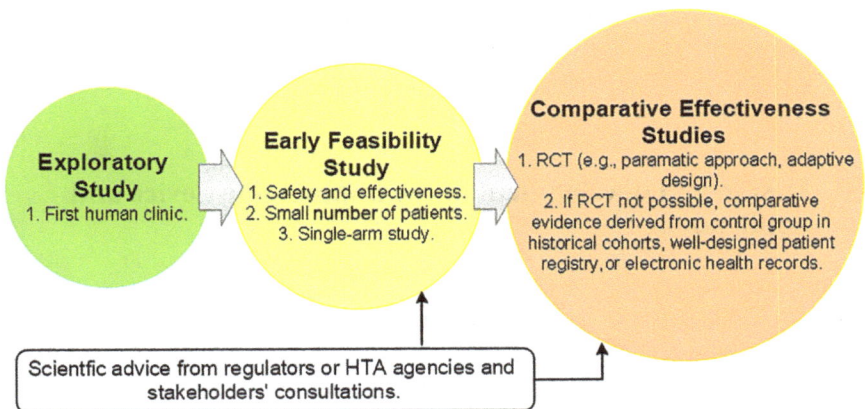

Figure 1.1   Clinical evidence generation in the clinical, pre-market stage

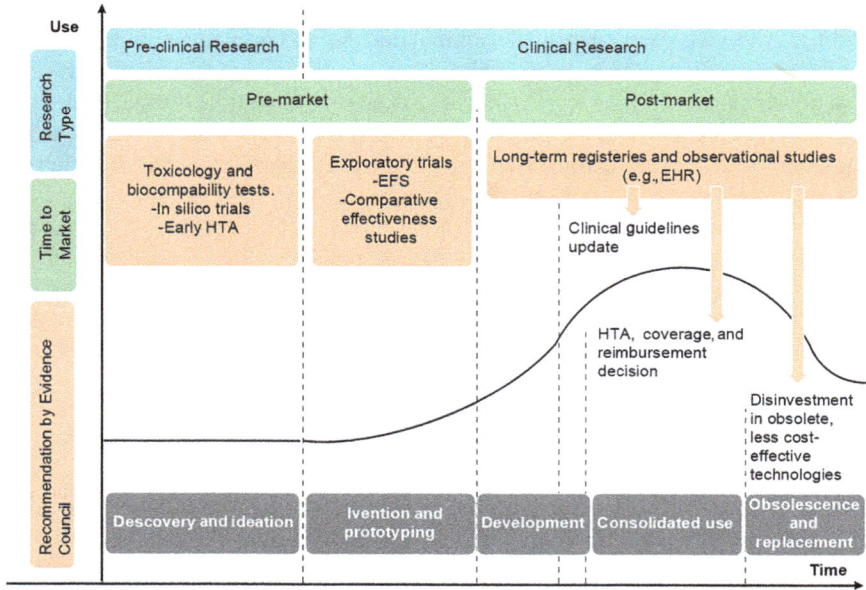

Figure 1.2   The life cycle evidence of a high-risk implantable device

contexts [87]. This modeling provides valuable insights into the fundamental properties of medical devices and how they interact within the patient's body. The FDA has approved the use of computational simulation to augment device evaluations by complementing bench testing, forecasting animal and clinical results, designing clinical trials, and examining efficacy, effectiveness, and safety [68,94]. However, before research begins, this software or these models must undergo verification and validation [68,94]. The specifics of applying computational simulation for device design and evaluation are detailed in Section 1.4. Clinical evidence is approached through exploratory methods to address potential issues that may impact the progression or halt the development of the device [87].

The life cycle for an implantable device, from design to clinic, is shown in Figure 1.2 [87].

1.4   Design and development of the implantable device and applications

The healthcare sector's capacity to deliver safe, high-quality patient care hinges on the judicious integration of effective design and clinically efficacious medical devices. A medical device cannot function optimally unless its design is tailored to its specific usage context. It is crucial to consider the technical requirements of healthcare settings alongside the varying contexts in which healthcare professionals, patients and caregivers utilize these devices. These tools must seamlessly fit into the existing workflows of healthcare workers, enhancing care delivery while accommodating the daily needs and preferences of patients and their caregivers [64]. The term "medical devices" encompasses an extremely diverse spectrum, spanning from simple items like adhesive bandages to intricate machines like heart-lung devices, which play a critical role in contemporary surgical procedures. This wide-ranging category illustrates the diversity in design, as functionality, reliability, and usability vary significantly across different device types. Simplistic devices prioritize safety, user-friendliness, and wide accessibility, allowing patients to use them independently. Conversely, complex machines designed for operation by skilled medical staff require specialized designs to ensure accuracy, minimize errors, and maintain consistent performance within stringent clinical parameters. Consequently, designing medical devices demands a holistic, user-focused approach that accounts for situational variability. Medical devices only contribute to safe and effective care by addressing the needs of users and the environmental context. Over the past decade, various initiatives have been introduced to enhance the safety of medical equipment. Efforts have been made to bolster the safety and reliability of medical devices. Established in the United Kingdom in 2001, the national patient safety agency (NPSA) aims to enhance patient care within the national health service (NHS) and enforce safety protocols by systematically and comprehensively reporting incidents of harm and "near misses" in healthcare settings. These preventive measures form part of a broader effort in healthcare to develop safer, more user-centric technologies. By focusing on the enrichment of medical devices' physical design, including their usability and interaction with both patients and healthcare professionals, these programs represent a significant step forward. Ultimately, this could lead to a more dependable healthcare environment, reducing the likelihood of device-related errors and fostering a culture of safety and continuous improvement in medical care [13].

In recent times, it has been acknowledged that effective design crucially enhances patient safety, prompting extensive research on the usability of particular medical devices, notably infusion pumps [64]. Creating a truly user-centric medical device necessitates addressing a wide array of considerations. While clinical needs, eliminating human error, and boosting patient safety are key priorities; the scope of medical device design extends well beyond simply fulfilling user requirements [32,58,64]. Design efforts should be robust and align with the natural work flows and routines of healthcare personnel. For devices to function effectively and efficiently, they need to integrate seamlessly into existing workflows, allowing providers to concentrate on patient care without needing to adjust the device itself. Ideally, the device should be ready to use without any personal adaptation. Furthermore, a well-conceived device encourages sustained use, vital for adherence to treatment and intervention schedules, consequently improving patient outcomes. It is also essential to consider the broader system in which a device will operate, given the complexity of healthcare systems that involve various users across multiple settings – from emergency and acute care to home healthcare. Compounding this complexity are factors such as remote purchasing, variations in clinical conditions, health statuses, and disabilities among clients. These differences, coupled with the unique characteristics of patients and providers necessitate device designs that can adapt to diverse settings and user profiles [64].

Ultimately, while focusing on the technical and clinical aspects of medical device design is crucial, it is equally important to address the diverse contexts and user interactions in healthcare. Tackling these complexities significantly enhances medical devices' capability to facilitate safe, efficient, and effective patient care across various healthcare environments [98].

Understanding the multifaceted requirements of users can pose a challenge for developers, even for those possessing clinical expertise, as user needs extend beyond mere clinical efficacy. Consideration of the diverse demands of a broad spectrum of users, encompassing doctors, nurses, technicians, maintenance personnel, patients, and caregivers, is essential for each specific piece of medical equipment. Typically, the user base of the device encompasses a wide array of operators in addition to patients, who are the recipients of the care facilitated by the technology. The diverse range of operators involved may include both clinical and nonclinical staff, trained professionals designated as intended users, and additional caregivers who may interact with the device without being its primary operators [64]. Unique to the realm of medical device development is the necessity to assimilate and address the varied user needs throughout the design process. In this regard, it diverges from other domains of ergonomic design, where the user base often exhibits greater homogeneity. Ensuring usability and efficiency for diverse categories of intended users with varying needs and skill levels is among the most critical steps in the creation of medical devices that provide safe, effective, and efficient care across different health management settings. The evolution of implanted medical devices spans several millennia. Nevertheless, significant advancements, commencing with the first implantation of a pacemaker in 1958, have been pivotal in the progression of the contemporary implantable device industry. The collaborative efforts among university researchers, industry leaders, and the medical community facilitated the development of numerous new implants designed to address a wide array of medical conditions. This collaboration fostered the establishment of a distinct market for implantable devices while simultaneously promoting the advancement of various types of devices. According to the Freedonia Group, US demand for IMDs was projected to grow at an annual rate of 7.7% to 52 billion USD by 2015. Considering the crucial role these devices have assumed in modern medicine, coupled with ongoing scientific exploration of fundamental principles that enhance their efficacy and consequently their adoption, this forecast underscores the rapid expansion and high demand for such devices [65]. The IMD market is predominantly led by orthopedic implants, predominantly utilized for bone and joint replacement. Other key categories include cardiovascular implants, such as pacemakers, stents, and structural supports, alongside neurostimulators and drug delivery implants. These diverse device types are designed to address a broad spectrum of medical ailments and disorders, catering to a variety of therapeutic and diagnostic needs. The primary focus within this section will be implantable devices incorporating electrical components, some of which may operate with or without active electrical power. Examples of such devices include implanted pacemakers, cochlear implants, medication infusion pumps, pressure sensors, and various types of stimulators. These devices play a crucial role in monitoring physiological parameters pertinent to specific medical conditions or administering therapeutic interventions.

Special emphasis will be placed on implants utilizing advanced microtechnologies, including microelectromechanical systems (MEMS) [64]. Currently, numerous medical implants incorporate microfabricated components, with some relying entirely on microfabrication techniques. These encompass devices approved for clinical use, as well as those intended for research purposes. Recent innovations include cardiovascular defibrillators and pacemakers equipped with MEMS accelerometers to detect motion and body position, thereby enhancing their capacity to adapt to patient needs [64]. The subsequent chapter explores emerging technologies applied in IMDs.

1.5   Specific technologies in implantable device

1.5.1   CMOS technology

Metal-Oxide-Semiconductor (CMOS) technology is pivotal in the production of implanted devices due to its outstanding efficiency, minimal power requirements, and scalability. This technology enables the integration of complex electronic circuits into a single microchip, which is critical for developing compact, energy-efficient devices suitable for implantation in humans [41].

*   **Minimal power consumption:** CMOS circuits are incredibly power-efficient, consuming very little energy. This efficiency makes them especially suitable for implanted devices that operate on alternative energy sources, such as energy harvesting methods, or have limited battery capacity. The low-power demand is crucial for minimizing the need for replacements or frequent recharging of the device, enhancing patient comfort, and extending the lifespan of the implant [16].
*   **High integration and miniaturization:** CMOS technology permits the integration of multiple functionalities, such as data transmission, signal processing, and sensing, onto a single chip. This high degree of functional integration allows implantable devices to be made significantly smaller, less intrusive, and more comfortable for patients [50].
*   **Scalability and cost-effectiveness:** Known for its scalability and cost-effectiveness, CMOS technology is extensively used in the semiconductor industry. Its widespread availability helps lower the production costs of implantable devices, making advanced medical technology more affordable and accessible to a broader population [26].
*   **Wireless communication:** CMOS technology supports wireless communication modules that are essential for transmitting data from implanted devices to external monitoring systems. This wireless capability is integral to modern healthcare, offering remote diagnostics and real-time patient monitoring [38,50,69].

CMOS technology plays a critical role in the advancement of implantable device technology by enabling key features such as wireless connectivity, miniaturization, power efficiency, and broad functionality, all of which improve the reliability and effectiveness of these medical devices. The CMOS industry remains crucial for implantable devices due to their capabilities in achieving low-power consumption, reduced size, and cost-effectiveness.

1.5.2   MEMS technology

MEMS technology has emerged as a key component in the development of IMDs due to its ability to reduce size while maintaining precision and capability. Essentially, MEMS integrates mechanical components, sensors, actuators, and electronics on a miniature scale to create adaptable parts for use within the human body. This technology enhances both diagnostic and therapeutic functions in critical applications that rely on implantable device design. MEMS technology facilitates the merging of electronic and mechanical components at a microscopic level, enabling capabilities like physical sensing and actuation. As a result, MEMS are ideally suited for applications such as chemical sensors and include features like micropumps for drug delivery. MEMS can interact dynamically with their surroundings through mechanical actions or by sensing mechanical forces and converting them into electrical signals for monitoring purposes. However, when compared to CMOS technology, MEMS systems lack the ability to process complex signals within a compact area [18,72,77,88].

- **Miniaturization and power efficiency:** MEMS's capability to function with minimal power and be reduced to microsized components is particularly advantageous for sensing applications requiring extended observation over time. Nonetheless, power efficiency is dictated by the type of mechanical interactions, with active MEMS components potentially consuming more power than passive CMOS circuits [2,35,49,85,103].
- **Integration capabilities:** Due to the limited processing power of MEMS, these components are often integrated with CMOS circuits for handling data. Typically, MEMS units are developed as standalone modules that interface with CMOS circuits to process sensor outputs. For instance, an MEMS pressure sensor might use CMOS circuitry for processing and transmitting sensor data to an external system. Meanwhile, CMOS technology allows for the integration of sophisticated electronic circuits on a single chip, facilitating the consolidation of multiple operations such as data processing, wireless communication, and memory storage on one platform. Moreover, CMOS technology is compatible with other technologies, like MEMS, frequently acting as the computational core that drives the mechanical sensing functions of MEMS [42,74,79].
- **Application in wireless communication:** In wireless communication, MEMS devices generally require integration with CMOS circuitry. MEMS sensors, such as those measuring pressure or detecting acceleration, capture environmental data, but rely on CMOS-based modules to wirelessly transmit this information to other systems [54,73,76,82].
- **Biocompatibility and structural composition:** Using biocompatible materials such as silicon, certain polymers, or biofriendly metals, MEMS devices are suitable for safe, long-term implantation. Due to their mechanical nature, MEMS can also be coated with materials that increase durability and biocompatibility in body environments. Although CMOS components can be made biocompatible, they are typically enclosed within protective casings to prevent contact with

the bodily fluids. As CMOS uses silicon-based transistors and electrical components, it is prone to corrosion in physiological environments if unprotected, making its direct exposure is more limited compared to MEMS [36,43,52,95].

CardioMEMS, Inc.'s EndoSure AAA Wireless Pressure Measurement System marked the first FDA-approved MEMS implant. This sensor is designed to monitor internal pressure within an aneurysm during endovascular repair of abdominal aortic aneurysms. Similarly, ISSYS Sensing Systems, Inc. is working on developing wireless microfabricated pressure sensors with applications across various conditions like traumatic brain injury, pulmonary edema, congestive heart failure, and hydrocephalus. Their Titan Wireless Implantable Hemodynamic Monitor (IHM) device is currently in clinical trials, assessing pressure within the left atrium or ventricle. These innovations underscore the increasing role of microfabrication in enhancing medical implant precision and functionality [64]. Fundamentally, MEMS excels in physical sensing and actuation, managing interactions with internal environmental or mechanical forces, while CMOS is optimized for data processing, transmission, and storage, due to its high integration capabilities and low energy consumption. In advanced implantable devices, both technologies complement each other: MEMS captures and responds to physical signals, whereas CMOS circuits adeptly process, transmit, and control data, supporting advanced medical implants.

1.5.3   Flexible electronics technology

Flexible electronics technology is used in the production of implanted biomedical devices to improve their function, comfort, and adaptability. Flexible electronic devices are typically crafted from elastic, bendable materials, allowing them to seamlessly conform to bodily shapes and movements. This design minimizes tension and irritation on surrounding tissues [17,40,57,97].

Utilizing comfortable biosensors facilitates continuous monitoring as these flexible sensors can directly adhere to tissue surfaces. For example, flexible cardiac sensors can be mounted on the heart's surface to track electrical signals, blood flow, or pressure changes. Cortical electrocorticography (ECoG) electrodes made from flexible materials conform to the brain cortex for neural activity recording and epilepsy monitoring. The adaptability and flexibility of these sensors ensure sustained contact with dynamic physiological environments, yielding more accurate and continuous data. Their enhanced biocompatibility and minimal tissue interference make them suitable for extended implantation [53,62,63]. Beneath the skin, flexible electronic devices can assess physiological markers such as lactate and glucose levels. These sensors can be implanted subcutaneously to monitor levels in real-time, provide data to other devices, or directly interface with drug delivery systems to adjust dosage based on sensor feedback. Rigid devices in the implantation area could cause discomfort; flexible sensors can move with the skin and muscles. Flexible glucose sensors enable diabetic patients to monitor their levels accurately while maintaining comfort in their daily activities [12,47,110]. Many neural stimulation and recording technologies, including brain–computer interfaces (BCIs) and neuromodulation technology, use flexible electrodes. Flexible neural electrodes, which can be inserted directly into the spinal cord or peripheral nerves, ensure accurate stimulation and signal capture. The flexibility and small size of these electrodes allow them to better integrate with the neural tissue architecture, reducing tissue damage and inflammatory reactions often caused by rigid electrodes. This approach can benefit epilepsy treatment, Parkinson's disease, and chronic pain management [84,96,104,108]. Implantable cardiac monitoring and pacemaker devices, such as flexible pacemaker electrodes and internal electrocardiogram (ECG) monitors, also greatly benefit from flexible electronics. By conforming closely to the heart surface, flexible pacemaker electrodes reduce tissue friction and discomfort associated with traditional rigid devices. Flexible cardiac devices enable noninvasive, precise long-term monitoring by adapting to the heart's expansion and contraction. These flexible pacemaker electrodes are more biocompatible than their rigid counterparts, reducing rejection and inflammation [102,106,107]. Flexible optoelectronic devices can attach directly to the retina, aiding visually impaired individuals in regaining partial vision through optical sensors and retinal prostheses. Flexible retinal implants can respond to external light stimuli and transmit signals to the optic nerve to restore visual information. Their comfort and durability are enhanced, minimizing retinal damage and making them suitable for sensitive regions requiring precise implantation. Moreover, these devices offer remarkable biocompatibility and flexibility, facilitating a more natural restoration of vision [40,57,97].

1.5.4   Nanotechnology

Nanotechnology plays a crucial role in enhancing the sensitivity, precision, and biocompatibility of biomedical devices on a nanoscale, making it an integral part of implantable healthcare devices [25]. These implantable technologies can significantly boost diagnostic and therapeutic capabilities by efficiently interacting with biological systems, thanks to nanoparticles, nanostructures, and nanomaterials [3,34]. Notable applications of technologies in the implantable devices include [61,81]: Nanosensors, due to their ability to detect biomarkers at extremely low concentrations, can facilitate the early diagnosis and monitoring of diseases like cancer, cardiovascular ailments, and infections. Integrated nanosensors in implantable devices can monitor physiological changes in real-time by measuring pH levels, glucose, or specific cancer markers in body fluids. The heightened sensitivity afforded by the large surface-to-volume ratio of nanostructures enables nanosensors to identify trace amounts of biomarkers. This capability is vital for early intervention and improving patient outcomes. Furthermore, nanosensors can be integrated with wireless communication modules for data transmission and constant monitoring [37,56]. Implantable devices also employ nanoparticles as drug carriers, enabling targeted delivery of medication to specific cells or tissues. For instance, chemotherapy drugs can be directly administered to tumor sites using drug-loaded nanoparticles in implantable devices, thereby minimizing systemic side effects while enhancing therapeutic impact. By directing drugs to specific cells or tissues, nanoparticles help to minimize off-target effects and improve treatment precision. Nanoparticles provide sustained drug release, benefiting chronic conditions that need continuous medication administration. This approach also increases patient comfort and adherence by reducing the frequency of drug delivery [46,101]. Implantable devices enhanced with nanocoatings display improved biocompatibility and reduced immune responses [11,14]. Coating implants with nanomaterials like titanium dioxide or silver nanoparticles enhances their safety for long-term use by preventing bacterial colonization, minimizing infection, and reducing inflammation. Over time, biofouling – the accumulation of proteins and cells on device surfaces – can impair functionality. Nanocoatings effectively counteract this. Antimicrobial nanocoatings lower infection risks from implants, especially in areas such as orthopedic devices, stents, and prosthetics. Customized nanocoatings that promote tissue integration can lead to better healing and device stability [11,14].

In regenerative medicine, fields like bone and nerve tissue engineering benefit from nanostructured scaffolds [75,90]. These scaffolds, either incorporated into implantable devices or applied as nanostructured surfaces, enhance tissue regeneration by promoting cell adhesion and proliferation. Commonly employed in wound healing, neuron regeneration, and bone implants, these scaffolds mimic the extracellular matrix, fostering an environment conducive to cell attachment, growth, and differentiation. For optimal long-term stability and functionality, rapid and effective tissue integration with the implant is crucial. Moreover, nanostructures aid in the controlled degradation of biodegradable implants, aligning with tissue repair processes [75,90,112]. For energy harvesting and storage in implantable devices, such as pacemakers and biosensors, materials like graphene and carbon nanotubes are under study. These devices can feature nanostructured supercapacitors or batteries to harness and store energy from movement or body heat, ensuring prolonged device operation. The outstanding energy storage capacity and electrical conductivity of nanomaterials make them ideal for devices demanding consistent and enduring power. Energy solutions rooted in nanotechnology can enable self-sustaining operation of implantable devices, reducing or even negating the need for frequent battery replacements, which is particularly beneficial for devices implanted in inaccessible locations [27,39]. Nanotechnology enhances implantable biomedical devices by offering more sensitive detection, precise drug delivery, improved biocompatibility, regenerative capabilities, and increased energy storage. This enables the development of more patient-friendly, durable, and high-performance implants, paving the way for intelligent, targeted, and minimally invasive medical treatments. Consequently, nanoscale engineering of implantable devices can better interface with biological systems, leading to significant advancements in therapeutic and diagnostic applications [25].

1.5.5 Energy harvesting technology

Implanted biomedical device manufacturers highly esteem energy harvesting technology due to its ability to prolong battery life, reduce battery replacement frequency, and offer renewable energy sources. As these technologies progress, there is potential for implanted devices to operate independently by harnessing energy directly from the environment or body, particularly in scenarios where long-term implantation is required [31,44,86].

Thermoelectric energy harvesting: Thermoelectric generators are advantageous for supplying reliable power to low-power implantable devices by converting body heat into electrical energy. These generators are particularly suitable for devices located in areas with stable temperature gradients, such as beneath the skin or near major blood vessels, where heat is consistently present. This noninvasive energy source draws on the natural temperature difference between body heat and the surrounding air, offering a dependable power solution. It is especially well-suited for powering devices like pacemakers, glucose monitors, or sensors that need continuous operation without high power demands. Additionally, thermoelectric systems can minimize the frequency of replacement procedures due to their long-lasting performance and low maintenance costs [30,31].

Piezoelectric energy harvesting: Piezoelectric materials generate electricity when subject to mechanical stress or motion. Implantable devices can leverage piezoelectric energy conversion by utilizing energy from natural physiological movements such as respiration, joint activities, and heartbeats. For instance, the mechanical energy generated by heartbeats can be transformed into electrical energy by embedding piezoelectric elements in a pacemaker. This demonstrates that piezoelectric energy harvesting is particularly effective under dynamic circumstances, making implantable devices in constantly moving locations more efficient. Devices like neural stimulators and defibrillators, which need rapid power bursts, could gain significant advantages. Furthermore, using various device designs, comfort is maintained, and piezoelectric materials remain biocompatible [70,91].

Electromagnetic energy harvesting: Induction serves as a method in electromagnetic energy harvesting that generates electricity through the interaction of coils and magnets. This technique has the capability to convert human movements or even the flow of blood into electrical energy for IMDs. For example, compact electromagnetic generators can be used in cardiovascular implants to harness the kinetic energy from heartbeats and blood circulation. Electromagnetic energy harvesting benefits devices requiring moderate to high power, like certain cardiovascular or motorized implants, by providing greater energy output compared to other methods. The technology can be adapted to a variety of frequencies and performs efficiently under highly dynamic conditions, aligning with the body's inherent movement patterns. Its versatility allows it to be applied across different environments and applications [15,33,80].

RF energy harvesting: Radio frequency (RF) energy harvesting involves collecting power from external RF sources such as Wi-Fi, mobile signals, or dedicated RF transmitters. Devices implanted with RF energy harvesters can capture and store these signals, utilizing them to power low-energy components like communication modules and biosensors. This technology is particularly advantageous for wirelessly transmitting health data to external monitoring systems. RF energy harvesting facilitates wireless recharging for devices requiring intermittent power for data transmission, eliminating the need for internal batteries and thereby reducing the device's size and complexity. Additionally, it allows for remote recharging, minimizing the necessity for invasive procedures to update or replace the device [24,29].

Energy generation via biochemical processes: Biochemical energy generation, known as biofuel cells, extracts energy from naturally occurring bodily fluids through chemical reactions with compounds like glucose and oxygen. This process mirrors the body's natural energy acquisition and can be applied to implantable devices to harness chemical energy. For example, glucose biofuel cells are particularly suited for continuous glucose monitoring devices because they convert blood glucose into electrical power. The reliability of biochemical energy generation for long-term implantable devices lies in its use of the body's inherent chemistry to naturally supply a stable, biocompatible energy source. This method supports implantable devices requiring constant, low-power usage without external recharging. Biofuel cells utilizing biochemical substrates from the body are sustainable and reduce the necessity for follow-up interventions or upkeep [4,55,111].

Photovoltaic energy harvesting: Photovoltaic (PV) energy harvesting involves converting light into electrical power, making it a practical option for devices implanted near the skin or in areas accessible to light, such as through transdermal illumination. PV cells can be integrated into devices like pacemakers or biosensors, allowing them to convert both directed and ambient light into energy to power low-energy components. This system offers a sustainable and noninvasive power source that can be replenished externally by exposure to light. This approach is particularly advantageous for subcutaneous implants, as it allows patients or healthcare providers to "charge" the implant by simply shining light on the skin. Even when light is limited, PV cells can couple with energy storage solutions to ensure continuous power availability [109].

Energy harvesting technologies are pivotal in the advancement of implanted biomedical devices, as they reduce the dependency on traditional batteries and facilitate sustainable functioning. Whether these devices are deeply embedded in the body or situated near the skin, the following energy harvesting methods – thermoelectric, piezoelectric, electromagnetic, radio frequency, biochemical, and photovoltaic – each offer distinct benefits tailored to specific implantation contexts. By harnessing motion, body heat, chemical reactions, radio waves, and light, these technologies extend the operational lifespan of implantable devices, minimize maintenance requirements, and enhance patient comfort. In particular, PV energy serves as a sustainable and user-friendly option for subcutaneous devices, proving to be a valuable addition to the collection of energy sources that power self-sustaining, long-lasting medical implants [31,44,86].

1.6 Quality and risk regulations in implantable device design

During the initial phase of developing IMDs, fewer technical and clinical validations were necessary compared to current standards. This was because the first-generation devices, such as artificial hips or cardiovascular stents, had simpler and more uniform features, making them less complex to implant [7]. However, as implantable systems began to incorporate more complex functions since the 1920s, greater safety and quality requirements were established to reduce risks within the human body. A notable advancement during this period was the cardiac pacemaker, followed soon by more advanced cardioverters/defibrillators. These devices have contributed to saving millions of lives and have also generated substantial commercial gains [7]. In 2001, the NPSA was established.

Medical devices are categorized into various classes and categories, according to distinct regulations that are based on particular stipulations. Nonetheless, several critical factors affect the classification of medical devices [64]:

- The potential risk to patients due to device application or malfunction.
- The length of time the device remains in contact with or is implanted in the patient's body.
- The degree of invasiveness.
- Local versus systemic effects.

Depending on the level of risk, the FDA classifies medical devices into three levels. The detailed breakdown of these classifications, along with relevant regulatory requirements, is provided in Table 1.1.

1.6.1   Importance of regulatory pathways

The reliability, effectiveness, and standard of IMDs heavily rely on regulatory processes. Regulatory bodies play a crucial role in establishing stringent standards for the design, testing, approval, and oversight of these devices, due to the potential hazards associated with equipment that interacts directly with the human body. Implanted devices such as pacemakers, insulin pumps, and neurostimulators engage with essential bodily functions, and any failure could lead to severe impacts on patients. As a result, regulatory pathways mandate a comprehensive evaluation process before devices are introduced to the market to mitigate these risks.

1.6.2   Regulatory agencies' function in safety and effectiveness

In their respective regions, organizations such as the European Medicines Agency (EMA) and the U.S. FDA hold the responsibility for setting and enforcing regulations for medical devices. In the United States, the FDA's Center for Devices and Radiological Health (CDRH) manages the regulation of these devices by categorizing

Table 1.1   FDA classification of medical device, regulatory requirement and example devices

Classification	Risk	FDA regulatory pathway	Examples
Class I.	Low	Most exempt from regulatory process	Bandages, Wheelchair, Massage bed
Class II.	Low to medium	Most require 510(k)	Contact lens, Bone screw, Ultrasound machine
Class II. with control	Medium to high	Most require 510(k)	Glucose drug pump, Distal stent, Dialysis devices
Class III.	High	Most require PMA	Coronary stent, hip replacement, invasive neuron stimulator

them according to risk level and determining the appropriate approval path, whether that be the de novo process for innovative devices, the 510(k) clearance, or the pre-market approval (PMA) procedure. In the European Union, medical devices are regulated under the medical device regulation (MDR). Devices compliant with the EU's health, safety, and environmental protection criteria must bear the CE (Conformité Européenne, the French for European conformity) mark.

1.6.3   The global regulatory environment

Each country implements its unique framework and guidelines to ensure the safety and efficacy of medical devices, creating a varied regulatory landscape. Along with the FDA in the United States and the CE marking in the European Union, nations such as Japan, Canada, and Australia have established their own regulatory bodies and processes. For instance, Canada's Health Canada and Australia's Therapeutic Goods Administration (TGA) enforce stringent safety standards, similar to the rigorous regulations set by Japan's Pharmaceuticals and Medical Devices Agency (PMDA) that align closely with those of the FDA. The globalization of the medical device industry has led to initiatives aiming to harmonize the standards of different regulatory authorities. Organizations such as the International Medical Device Regulators Forum (IMDRF) work toward simplifying regulations worldwide and promoting device innovation without compromising safety.

1.6.4   FDA and CE marking: important approval routes

To access the US and European markets, manufacturers often must obtain FDA approval or CE marking. These regulatory pathways involve comprehensive evaluations of a device's clinical and technical performance. The FDA's requirements vary based on the device's classification: Class I (low-risk) devices need minimal regulatory oversight, while Class III (high-risk) devices are subject to an extensive PMA process. In Europe, CE marking involves a conformity assessment by a Notified Body that verifies the device meets the essential safety and performance requirements of the MDR.

While the FDA and CE marking follow distinct methodologies, both strive to ensure devices are safe for patients and effective for their intended applications. The EU MDR prioritizes post-market surveillance (PMS) and continual monitoring more than the FDA, which focuses on comprehensive upfront evaluations, including clinical trials for high-risk devices. Together, these strategies aim to create a comprehensive framework for maintaining device safety throughout the entire product life cycle.

1.6.5   Risk management and quality control

Regulatory mechanisms are crafted to address the hazards associated with implanted devices by enforcing strict standards for risk assessment and quality control. Agencies like the FDA and MDR mandate that manufacturers adopt risk management strategies in line with international norms such as ISO 14971, which focuses on the risk management of medical devices. These strategies entail identifying potential hazards, assessing the risks associated, and taking control measures to mitigate them. Moreover, manufacturers must establish quality management systems (QMS) compliant with standards like ISO 13485 to ensure uniformity in design, production, and quality control. Given that implanted devices remain in the body for extended periods and potentially perform vital life-sustaining functions, the adoption of risk management and quality assurance measures is crucial. For instance, pacemakers require rigorous testing to ensure battery longevity and resistance to electromagnetic interference. By following specific regulatory guidelines, manufacturers can effectively handle risks, maintain product quality, and demonstrate their dedication to patient safety.

1.6.6   Ensuring post-market surveillance

Regulatory frameworks incorporate comprehensive PMS systems alongside pre-market authorization. Once a device is in use, PMS becomes essential for monitoring its effectiveness, safety, and overall performance. Both the FDA and the EU MDR require robust PMS mechanisms, obligating manufacturers to consistently collect and analyze real-world data (RWD) on device functionality. While the EUDAMED database compiles post-market information across the European Union, the FDA's MDR system and MAUDE (Manufacturer and User Facility Device Experience) database in the United States facilitate the tracking of adverse events. By employing PMS, regulators and manufacturers can identify and address any unforeseen issues that may arise during the product's life cycle. For example, negative incidents involving patient complications or device malfunctions might lead to safety notices, product recalls, or design modifications. Ultimately, this continuous monitoring ensures patient safety and maintains public trust in medical technologies by safeguarding the efficacy and safety of devices.

1.6.6.1   The role of regulatory pathways in fostering innovation

Although primary regulatory pathways have been principally focused on safety and efficacy concerns, they significantly encourage innovation within the medical device sector. Clearly defined regulatory standards provide a critical framework necessary for the introduction of innovative products that are both safe and effective. For instance, the FDA's Breakthrough Device Program facilitates the expedited development and assessment of technologies that provide superior treatments or diagnoses for severe, life-threatening, or permanently debilitating conditions. Regulatory agencies also support technological advancement by collaborating with manufacturers, offering guidance on compliance with regulatory standards. Such collaborative efforts ensure the industry continuously evolves to improve healthcare while adhering to stringent patient safety criteria. In the design and manufacture of implantable devices, the importance of regulatory pathways cannot be understated. Through pre-market assessments, risk management strategies, and PMS, these measures uphold safety, quality, and efficacy. By enforcing compliance with FDA regulations, CE marking, and various international standards, risks are minimized, thereby safeguarding patient health. This regulatory framework fosters innovation, enabling manufacturers to develop advanced medical devices that enhance healthcare quality for patients.

Consequently, regulatory pathways are crucial for the appropriate development of technologies intended for IMDs.

1.6.6.2   Challenges in PMS of implantable devices

IMDs present distinct challenges for PMS in contrast to pharmaceuticals. The heterogeneity and intricacy of device designs, alongside the iterative nature of device development and the learning curve required for clinicians to adopt new technology, further complicate consistent long-term safety monitoring [92]. Unlike pharmaceuticals, which have stable formulations, medical devices, ranging from pacemakers to orthopedic implants, often exhibit brief product life cycles, leading to the frequent replacement of older models with newer ones [92]. This rapid cycle of innovation has the potential to outstrip the capacity to collect comprehensive safety data for each version. In the meantime, older implanted devices may remain within patients for several years or even decades, potentially leading to the manifestation of issues only after extended utilization. A primary challenge involves the prolonged implantation duration and the latency of certain device-related problems. Devices such as pacemakers, artificial joints, or cochlear implants are designed to endure for numerous years within the human body; consequently, specific failure modes or adverse effects (e.g., material wear, battery degradation) may surface only over an extended period. Observing a rare yet serious complication necessitates the monitoring of a sufficiently large cohort of patients over an adequate duration of time. Several pertinent questions emerge: Do existing surveillance systems enroll a sufficient number of patients to statistically detect a suboptimal device performance, and are outcomes monitored at appropriate intervals [5]? Given that thousands of devices necessitate monitoring over many years to identify a safety signal, timely detection remains challenging. For instance, the significant failures observed in metal-on-metal hip prostheses only became apparent after the implants had been available on the market for years, as national joint registries began consolidating long-term outcomes [5]. This scenario exemplifies how the prolonged lifespan of implants and the occurrence of rare events can delay the identification of device-related issues.

The tracking of device failures is significantly impeded by underreporting and data deficiencies. In practical terms, many adverse events associated with implants are neither promptly nor systematically reported. In the United States, manufacturers are legally obligated to report device-related fatalities, serious injuries, and malfunctions within 30 days, with these reports populating databases such as the FDA's MAUDE [28]. However, a substantial portion of incidents are either reported belatedly or not at all. A recent analysis of 4.4 million FDA reports from 2019 to 2022 revealed that only approximately 71% were submitted within the stipulated timeline; nearly 9% were reported more than six months late, including numerous cases involving patient fatalities [23,28]. Such reporting delays result in regulators and clinicians potentially remaining unaware of emerging safety concerns, thereby causing preventable patient harm [28]. Furthermore, passive adverse event reports frequently lack completeness or verification, which restricts their utility [93]. The issue of underreporting is exacerbated by the difficulty in recognizing or attributing device malfunctions; for example, if a patient's health progressively declines due to a malfunctioning implant, it may not be immediately clear that the device is the cause. Another significant challenge is the fragmentation of data and the lack of interoperability across surveillance systems. Post-market data on implants are sourced from various origins – manufacturers' complaint databases, hospital records, insurance claims, national registries, etc. – that are often isolated and incompatible. The exchange of PMS data between stakeholders is limited, with many PMS databases being offline, localized systems that do not interact with each other [71]. Hospitals and countries operate distinct reporting systems, rarely providing a cohesive global perspective. Additionally, privacy regulations and jurisdictional controls necessitate that health authorities oversee data exchanges, complicating the international aggregation of information [71]. This lack of interoperability can result in missed safety signals if they are spread across multiple data sets. For instance, a registry in one nation might identify an issue well ahead of others, and without rapid data sharing, global alerts are delayed. The situation concerning metal-on-metal hip implants, where the Australian registry identified high failure rates before other regions, highlights the necessity for improved international data linkage – an aspect acknowledged by subsequent analyses of registry efficacy [5]. The development of mechanisms to enable different PMS stakeholders – such as manufacturers, hospitals, and regulators– to share and consolidate data in real-time continues to be a pressing challenge.

Moreover, some outcomes of interest are not well represented by typical surveillance experiences. Registries and administrative databases typically capture definitive endpoints, such as revision surgeries, device removals, or mortality. However, patient-reported outcomes (PROs), such as pain relief, functional improvement, and quality of life, provide important information about implant performance. These relevant but weak signals often predate more obvious device failures and are infrequently collected in a comprehensive way. The challenge of collecting PRO measures as part of long-term device surveillance remains, as these data sources do not routinely capture this information [5]. As one analysis put it, the prospects for developing approaches to collect patient-reported symptoms or satisfaction data would require the creation of new infrastructure (such as periodic patient questionnaires or mobile health data collection) that goes beyond the typical scope of most PMS programs [5]. Until such patient-focused data are collected, subtle device issues that impact quality of life (but not yet clinically significant) may not be detected by regulatory authorities. Lastly, the nature of medical device use presents challenges for surveillance. Many implant outcomes depend on user technique and other contextual factors. For example, the success of a cardiac implantable electronic device depends on the accuracy of its placement and calibration by the implanting clinician; the outcome of surgical mesh procedures may depend on surgical skill and patient characteristics. It is difficult to differentiate whether an adverse outcome was due to device design, practitioner technique, or patient comorbidity. The learning curve for new implantable technology means that early complication rates may be higher as practitioners become more experienced [92]. This situation may mask a safety signal, and an important question is whether a problem is due to a real device issue or a temporary one attributable to the learning curve. These considerations make clear that attributing causality in PMS is more difficult than causality attribution in drug surveillance. These issues highlight the need for effective, comprehensive approaches to implantable device PMS to detect problems in field.

1.6.6.3   Current regulatory and technical measures

Regulators and industry have implemented several measures to strengthen PMS of implantable devices, both through mandatory reporting systems and proactive data collection. In the United States, the FDA's cornerstone is the MDR system, a passive surveillance framework established by regulation (21 CFR 803). Under this system, device manufacturers, importers, and user facilities are required to promptly report any device-related deaths, serious injuries, or malfunctions to the FDA [93]. These reports are aggregated in FDA databases (e.g., the MAUDE repository) and are reviewed for patterns that might indicate a safety concern [93]. The FDA also encourages healthcare professionals and patients to submit voluntary reports (via the MedWatch program) to capture problems that might otherwise go unreported [93]. While useful, the FDA acknowledges that the MDR process alone is insufficient – by its nature it may miss incidents or contain inaccuracies, so it constitutes only one component of a broader surveillance system [93].

In addition to passive reporting, the FDA has the ability to require on manufacturers to undertake proactive post-market study requirements. Under Section 522 of the Federal Food, Drug, and Cosmetic Act, the FDA is authorized to require PMS studies of certain specific devices (generally Class II or III implants only) when it is needed to obtain more rigorous data regarding device performance and safety under real-world use conditions. Such studies may consist of prospective cohort studies, data registries, or other surveillance approaches that manufacturers are required to conduct and report to the agency. Both the United States and other countries have improved their use of RWD to inform methods for device surveillance. RWD is defined as health-related data elements collected outside of a clinical trial setting, including electronic health records, insurance claims, and patient registries. The CDRH of FDA has supported the use of RWD/RWE (real-world evidence) for regulatory decision-making. One major effort in this area is the National Evaluation System for Health Technology (NEST), a partnership to use data collected from clinical registries, health systems, and claims databases to perform active surveillance of device safety [5]. By aggregating large data sources, NEST will aim to detect safety signals with near real-time precision and provide evidence about device performance across different demographic populations and longer time horizons. For instance, large integrated health systems have used registry and EHR data to detect implants with observed failure rates greater than expected, allowing for precocious interventions (e.g., large integrated health systems have detected certain implants associated with revision rates greater than expected; the data from this can be used by clinicians and regulators to make decisions.). Proactive surveillance based on RWE has the potential to reveal issues that may escape passive reporting, thus strengthening the PMS ecosystem as a whole [5].

Device registries are a cornerstone of RWD initiatives. Many at-risk devices that are implanted today have dedicated registries that collect outcomes. For example, within cardiology in the United States, the implantable cardioverter-defibrillator

(ICD) Registry collects data on implanted defibrillators, and international collaborations exist for the tracking of pacemakers and heart valves. Orthopedic implants are monitored by national joint registries (existing in the United Kingdom, Australia, Sweden and many others) that record revisions and comprehensive information on characteristics of the device models. Registries are successful; for example, the recent failures of the aforementioned metal-on-metal hip implants were detected through the analysis of data from registries that revealed unusually high revision rates [5]. Registries amass data on performance for almost all implanted devices within a given category and allow for more transparent and independent monitoring of outcomes for devices [5]. Regulators have begun to incorporate registry data into safety communications and policies. The FDA has supported the creation of coordinated registry networks (CRNs) that integrate the many databases of separate registries (typically including clinical registries and claims data) to get a fuller picture of device performance [5]. These initiatives represent a shift toward a more proactive PMS system in use today, one that might be described as a form of "technovigilance," highlighting the need for continuous, data-driven surveillance rather than a sole dependence on spontaneous reporting.

PMS in European Union has seen a significant improvement with the advent of new Medical Device Regulation (EU MDR 2017/745). As per EU MDR, manufacturers shall constitute an overall PMS system and develop a detailed plan for each device (Articles 83 and 84) to systematically collect and analyze data after marketing of the device. The regulation stipulates that an overall PMS system shall be put in place to develop a plan to proactively and systematically collect relevant data regarding relevant data regarding pertinent data regarding pertinent with regard to with regard to quality, performance and safety of the device throughout its life cycle [22]. The goal is to continuously reverify and revalidate the risk/benefit profile of the device using real-world usage data to aid risk management and product improvements [22]. Devices classified as higher risk (Class IIa, IIb, III) are required to provide a Periodic Safety Update Report (PSUR) at periodic intervals to designated notified bodies and regulators within prescribed limits of time (Article 86) for devices classified as lower-risk, a PMS Report is required (Article 85). These reports shall encompass adverse events, trends of nonserious incidents, results of any post-market studies, and any field safety corrective actions that have been taken. With the requirement of these reports, EU MDR has created a formal feedback loop that compels manufacturers to evaluate and take actions based on post-market data (Figure 1.2) instead of evaluating data on an ad-hoc basis [22]. In addition to improving PMS requirements, EU regulations have tightened vigilance requirements as well. Manufacturers are mandated to notify serious incidents to European authorities within a limited period (typically 15 days) and trend of minor incidents must be studied and evaluated and if necessary reported (Article 88 provisions regarding trend reporting). The EU is also moving ahead with development of EUDAMED, a comprehensive EU-wide database that will compile information with regard to devices, economic operators, their certified bodies, and vigilance data. Once fully implemented, EUDAMED will enable international information sharing with regard to performance and safety of devices.

Regulatory entities worldwide are increasingly collaborating to enhance PMS. Through organizations such as the IMDRF, principles aimed at enhancing PMS and data exchange are being standardized. For instance, the IMDRF has provided guidance on standard adverse event terminology and coding to improve the uniformity of reporting across various jurisdictions. Furthermore, there exist collaborative networks for the international exchange of device incident reports (such as the National Competent Authority Report exchange in the EU and analogous programs in other regions), ensuring that hazards identified in one area can promptly alert others. From a technical perspective, a significant innovation is the implementation of Unique Device Identification (UDI) systems. The FDA established UDI requirements between 2013 and 2014, and the EU MDR similarly mandates UDIs on medical devices and implants. A UDI is a code, presented both in human-readable and barcode formats, that distinctly identifies the device model and its production details. When UDIs are recorded in patient records and registries, they facilitate unambiguous referencing of the same device across various data sources. This greatly enhances traceability; for instance, in the event of a recall of a pacemaker model, UDIs enable healthcare institutions to swiftly determine which patients received that specific model. Additionally, UDIs facilitate the linkage of previously unrelated data sets. As noted by the FDA, UDIs allow the integration of information from Electronic Health Records (EHRs), clinical databases, and claims data that was previously "untapped," thereby unlocking new prospects for comprehensive device performance monitoring [92]. In conclusion, contemporary PMS strategies for implants integrate obligatory reporting systems (such as the FDA's MDR, EU vigilance reports) with the proactive collection of RWE (such as registries and post-approval studies), supported by new instruments like UDIs and coordinated data networks that augment the detection of safety signals.

1.6.6.4   Case studies: device recalls and lessons learned

Real-world case studies of implantable device failures highlight both the importance of rigorous PMS and areas where surveillance has fallen short. One representative example is the recall of certain cardiac implantable devices (pacemakers and defibrillators) due to premature battery failure. In 2016, St. Jude Medical (now part of Abbott) recalled hundreds of thousands of ICDs and cardiac resynchronization therapy devices after reports that their lithium-based batteries could deplete rapidly and without warning [19]. An investigation found that lithium clusters could form within the battery, causing a short-circuit and sudden battery failure [8,67]. This defect had dire consequences: if an ICD battery died unexpectedly, the device could no longer deliver life-saving shocks or pacing when needed, leading to patient deaths in at least two reported cases [19]. The recall was classified by the FDA as Class I (the most serious type, indicating a reasonable chance of serious injury or death). A critical lesson from this case was the need for timely action and communication once a problem is suspected. In this instance, there were critiques of how the manufacturer handled the initial field notifications – the FDA later issued a warning letter noting that some affected devices continued to be shipped and implanted even after the issue was known [19]. This underscores that effective PMS is not just about detection of issues, but also about the efficiency of the response (hazard communication, device correction or withdrawal) once a red flag is raised.

From the St. Jude battery recall, several improvements were spurred in industry practice. Manufacturers developed battery monitoring algorithms and early-warning alerts for implants. Indeed, Abbott implemented a "Battery Performance Alert" system via a firmware update to give physicians early warning of unusual battery drain in patients' devices [8,20]. More broadly, the case illustrated the importance of collecting failure mode data and investigating root causes for device malfunctions. Post-market analysis of returned devices revealed the lithium cluster phenomenon, which was not detected in pre-market testing. This knowledge has informed better design and quality control for subsequent device batteries. Regulators, on their side, learned to enforce more stringent oversight on how recalls are managed, ensuring that all stakeholders (hospitals, patients) are promptly informed of risks. The pacemaker/ICD battery recall case also brings up the ethical consideration of prophylactic device replacement: surgeons and patients had to decide whether to explant and replace devices that had not yet failed. Such decisions weigh the risk of device failure against the risks of an invasive replacement procedure. In this case, patients were advised to closely monitor their device's battery status and elective replacement was considered for those at highest risk. The outcome emphasized a key lesson: post-market vigilance must continue throughout the device's service life, and contingency plans (like noninvasive device diagnostics or remote monitoring) are vital to manage risks in already-implanted devices.

Another instructive case study comes from cochlear implants, which are complex electronic implants used to restore the sense of hearing. In 2020, Advanced Bionics – one of the major cochlear implant manufacturers – issued a voluntary recall (field corrective action) for its HiRes Ultra and Ultra 3D cochlear implant models after numerous reports of device failure began to surface [66]. Patients with affected units experienced sudden degradation of hearing performance; investigation traced the issue to fluid ingress at the electrode feedthrough seal, which caused electrical shorting and loss of function [66]. Essentially, moisture was able to leak into the implant casing in some units, damaging the electronics. By early 2020, hundreds of such failures had been confirmed, and the company, in coordination with FDA regulators, recalled the affected batches. The recall required patients with faulty implants to undergo revision surgery to remove and replace the device [66] – a significant adverse outcome given the complexity of inner-ear surgery and the risk of cochlear damage with reimplantation.

The cochlear implant failures revealed a number of important PMS insights. First, they demonstrated the value of post-market device reliability monitoring by centers and surgeons. In one large cochlear implant program, clinicians diligently tracked the performance of implants and noted an unusual clustering of failures in specific recent models. A retrospective analysis at that center showed failure rates of the recalled implants exceeding 20% within about 2 years, far higher than expected [60]. Notably, this local data analysis indicated a problem perhaps sooner than the manufacturer's aggregate data did. It highlights how end users (hospitals and clinicians) play a critical role in surveillance by reporting and analyzing outcomes, complementing manufacturer reports. Second, this case underlined the human impact of device failures: patients, many of them children in the case of cochlear implants, had to undergo additional surgeries, experienced lapses in hearing during the period of failure, and endured emotional and physical stress. "Soft failures" (where a device malfunctions intermittently or partially) also became a focus, as they can be harder to detect and rely on careful clinical evaluation of a patient's symptoms [60]. The lessons learned pushed for better diagnostic tools to identify early signs of implant failure (Advanced Bionics, for example, advocated for the use of Electrical Field Imaging tests to detect failing electrodes early [60]). On the regulatory side, the cochlear implant recall drove home the importance of robust design validation and supplier quality management – the root cause, a faulty seal, was traced to a component issue that could potentially have been caught with more stringent testing. In response, manufacturers across the industry reviewed their component sourcing and testing processes for long-term hermeticity (especially since implants must function in a fluid environment for years). This case also underscores how recalls of life-changing devices require coordinated support: audiologists, surgeons, and the company had to work together to ensure patients received replacement devices and rehabilitative support as smoothly as possible.

But these are just examples. There are many, many more PMS case studies that have contributed to today's practices. When metal-on-metal hip implants were withdrawn worldwide in 2010–12, regulators learned that they should not be assuming that limited data during implantation is sufficient for predicting long-term device performance – those hips were approved on the basis of short-term data, and only after-market registries revealed the high rate of early failures and tissue damage and led to a worldwide withdrawal [5]. When implants for ICD leads (the wires that connect the ICD to the heart) failed, as in the recall of the Medtronic Sprint Fidelis lead in 2007, regulators learned that manufacturers should be doing diligent after-market follow-up on high-stress components; the recalled leads represented almost 268,000 implanted leads, and the recall demonstrated how the failure of a single component (the wire, which would sometimes fracture) can lead downstream consequences, and how manufacturers' practices in following up high-stress components and in notifying physicians of problem leads must improve; each device illustrates for regulators a single theme: that feedback from use is important. These experiences have led to the invention of better ways to monitor devices (from device registries to heart alerts from remote monitoring), and they have led to regulatory changes (for instance, requiring manufacturers to submit annual manufacturer reports of device performance to the FDA for certain implants, or to conduct post-approval studies addressing specific device risks identified during pre-market review).

1.6.6.5   Future trends in PMS: toward smarter, connected surveillance

The future of PMS for implantable devices will utilize new technologies and data to address some of its current shortcomings. The use of artificial intelligence (AI) and machine learning for signal detection is one promising area. The amount of PMS data is exploding (thousands of adverse event reports, registry entries, health records, etc.) and AI mining tools may help to find relevant data patterns that a human analyst might miss. For instance, natural language processing software could be used to mine health records' clinical narratives for device malfunctions or explantations that were not formally reported through the regulatory system. Early work has shown that AI searching platforms can retrieve and review relevant literature and incident reports surrounding device safety with the goal of predicting or detecting safety signals sooner [71]. In the near future, we will likely see regulators and manufacturers use machine learning models that analyze data streams for unusual activity (i.e., an early warning system). This is perhaps best conceptualized as a scenario in which an algorithm might generate a signal that more reports are being filed for a certain model of pacemakers' batteries depleting quickly, but well before this pattern was evident to human investigators (e.g., an increase in pacemaker replacements). AI could also be used to stratify patient risk: by analyzing patient characteristics and device data, algorithms could identify subpopulations that are at higher risk for an implant's failure, and therefore warrant preventive action [71].

A promising new development in the toolkit of PMS is wearable and remote monitoring. Today, many implanted devices are telemetric; for example, modern pacemakers and defibrillators can send device performance data and alarms from home monitors (or even the patient's own smartphone). This "always-on" remote monitoring can be applied toward broad surveillance goals. Rather than periodic clinical visits and passive adverse event reports, data from implants – such as battery lifetime, lead impedance, and detected physiological parameters – can be aggregated to discern potential device issues. Patient-owned wearable devices could supplement this "always-on" monitoring by tracking relevant physiological parameters or patient-related metrics that may indicate a problem with an implant. For example, a wearable heart rhythm monitor has the ability to detect when a pacemaker is failing to prevent heartbeats from pausing – in this case, the wearable would send an alert to the patient (and possibly his/her heart specialist) that the implanted device should be evaluated. In the future, we might envision a more integrated system of implantables and wearables: an implanted cardiac device could communicate with an external wristband smartwatch, which would send an alert to the patient and her/his heart specialist should a problem be detected. The system would have a beneficial impact on both the individual and population levels of care, yielding richer data for surveillance beyond the traditional passive clinical report system. As shown in one review, the growing prevalence of smartphones and wearables can enable direct patient reporting of a broad range of PRO measures and health status, which could be entered into device registries or device databases [5]. The major challenge will be keeping pace with the massive influx of data – ensuring data quality while respecting patient privacy – but if successful, it will broaden our traditional conception of PMS reporting well beyond current practices. In essence, the system may evolve to an "always-on" surveillance model, where every implanted device is part of a network that constantly reports its status.

The advent of emerging technologies is poised to revolutionize the manner in which data is disseminated and safeguarded in the context of PMS. Blockchain technology has been identified as a potential remedy to issues concerning the interoperability and trust of PMS data [71]. Within a blockchain-based PMS network, all stakeholders, including manufacturers, hospitals, and regulators, could function as nodes within a permissioned ledger, wherein reports concerning device performance or malfunctions are documented in a tamper-evident manner. This framework could guarantee the integrity of data (prohibiting records from being altered in an unauthorized manner) and facilitate the real-time sharing of critical safety information across international boundaries. A private, consortium blockchain for PMS could, for instance, automatically document an adverse event report submitted by a hospital, thereby rendering it simultaneously visible to the device manufacturer and regulators, strengthened by cryptographic assurances of the data's security and authenticity [71]. Furthermore, smart contracts, which represent self-executing code on the blockchain, might be deployed to initiate alerts or actions when specific conditions are satisfied (for example, if five independent hospitals register a similar incident with an implant, an automatic alert could be dispatched to all users of that device model). Although largely theoretical at present, pilot projects in this domain are currently in progress; they possess the potential to transcend the extant fragmentation of surveillance data by establishing a decentralized yet cohesive data exchange backbone.

Along with data technologies, one would expect the regulation of PMS to evolve to be more life cycle-oriented as well. PMS is being viewed less and less as an add-on to approval and more as a part of a device life cycle to produce evidence for iterative improvements. The concept of adaptive or continuous regulation is becoming more attractive: the approval of an implant may be based on relatively sparse data, and widespread approval or continued device presence may depend on post-market evidence collected from registries or adaptive studies post-market. This blurs the line between pre-market and during market and essentially puts the burden of demonstrating efficacy in real-world use on the device. We are seeing this in part with programs like the FDA's Breakthrough Devices, which are approved with expedited FDA action but then placed in a position to collect rigorous post-market data. In the extreme, future scenarios may have patients playing a greater role in surveillance, whether that's patient-reported registries or apps on smartphones that let users report their experiences with their own implant, thereby producing grain-by-grain safety data. Additionally, the natural evolution of UDIs will continue, and more advanced downstream innovations will occur as UDIs become ubiquitous. The application of big data analytics will be facilitated as the use of Unique Device Identifiers becomes more common; imagine data mining of electronic health records across a nation to understand patterns of a specific device's performance using its UDI as a common key distributed across databases. We could understand the subtle long-term effects (e.g., does an implantable deep brain stimulator lead to an increase in specific side effects after 7-10 years?) by correlating device identity with patient outcome in a nation-wide scale. Additionally, recalls and vigilance communication is aided by UDIs: prospectively, if there was a problem with a device being used in patients, a database could be queried to see who in the nation had that device, and perhaps automated notifications could be sent to their health care providers via electronic medical record. The combination of UDI infrastructure and advanced analytics like AI will allow for rapid action in response to issues with devices.

In other words, the future of PMS of IMDs will be bigger, more connected, and more proactive. Signals of safety will be detected more quickly with the help of AI and predictive algorithms. Always-on wearables and remote surveillance will capture today's gaps in surveillance, producing a steady stream of patient and device performance data. New regulations and blockchain and other data-sharing solutions may give rise to the networks of trusted, interoperable post-marketing surveillance systems that span the globe. Ultimately, the goal should be that all implantable devices in our bodies are being watched – and identified – as early as possible – while they are still able to be fixed – throughout their life – proactive PMS catching problems before they harm patients and enabling rapid interventions to maintain trust in implantable life-saving devices [5,71].

1.7   Challenges and opportunities of the implantable application

The development and use of implanted medical devices come with unique opportunities and hurdles. As these devices grow more intricate and advanced, they hold considerable potential for managing chronic illnesses and saving lives. However, substantial challenges relating to the market, design, and regulations need to be addressed for these devices to be successfully implemented.

1.7.1   Challenges in implantable device applications

The materials chosen for implantable devices must not trigger negative reactions in surrounding body tissues, such as inflammation or rejection. The choice of material is a crucial decision for designers, as every interaction between the body and the device can influence both safety and patient functionality. Sustaining the device's safety and effectiveness over extended durations poses an engineering challenge intensified by complex designs and material types. For long-lasting performance with minimal intervention, implantable devices often require dependable energy sources. Traditional batteries have limited lifespans and might necessitate frequent surgical replacements. Consequently, energy harvesting technologies like piezoelectric and thermoelectric devices, which harness energy from body movements or temperature variations, have evolved to address this issue. However, integrating these technologies into compact devices remains demanding. To minimize invasiveness and enhance patient comfort, implantable devices need to be miniaturized. This miniaturization introduces technical hurdles, as engineers must balance the need for functionality, longevity, and energy efficiency with the demand for reduced size. A significant technical challenge is embedding advanced features, such as wireless connectivity and sensor integration, into a compact form. Achieving regulatory approval is often intricate and time-consuming, particularly for Class III devices that pose a higher risk. These high-risk devices must undergo extensive testing and validation to comply with stringent safety standards set by authorities like the FDA and European MDR. The need for rigorous scrutiny of each new design or improvement can inhibit innovation, prolong time-to-market, and raise development costs. After deployment, implantable devices require regular monitoring to ensure safe and efficient operation. Manufacturers must establish robust PMS systems to track performance, manage recalls if necessary, and address emerging issues. This demands ongoing data collection and analysis, which can be complex and resource-demanding.

1.7.2 Opportunities in implantable device applications

Emerging biomaterials are offering new ways to boost biocompatibility and practicality. For example, bioresorbable materials naturally decompose over time, eliminating the necessity for surgical removal. Additionally, antimicrobial coatings play a critical role in preventing infection, which is a common issue with implants. These advancements pave the way for developing safer and more effective implants. Energy harvesting technologies, such as thermoelectric, piezoelectric, and radiofrequency energy harvesting, are presenting groundbreaking opportunities for implants. These methods allow devices to convert body heat, movement, or external radio frequencies into electrical energy, potentially supporting battery-free or self-sufficient electronics. This development is particularly promising for low-power devices like biosensors and pacemakers. Implantable devices can now benefit from wireless connectivity and integration with the Internet of Things, allowing for remote monitoring. This capability significantly enhances patient care by facilitating real-time health data exchange and prompt issue resolution. Moreover, wireless technology enables noninvasive data collection, reducing the need for intrusive follow-ups. AI and sophisticated data analytics integrated with implantable devices facilitate real-time data analysis and personalized treatments. AI algorithms can improve device maintenance by predicting potential failures, alerting healthcare providers, and adjusting treatment settings to optimize patient outcomes. Regulatory bodies are recognizing the pressing need for expedited pathways to bring essential implant technologies to market faster. Programs like the FDA's Breakthrough Devices Program aim to speed up the approval of innovative devices addressing unmet medical needs. These efforts reduce the time and cost associated with regulatory approval, allowing manufacturers to focus on developing and delivering novel treatments to patients more efficiently.

References

[1] S. Al-Jawadi, P. Capasso, and M. Sharma. The road to market implantable drug delivery systems: A review on US FDA's regulatory framework and quality control requirements. *Pharmaceutical Development and Technology*, 23:953–963, 2018.

[2] A.S. Algamili, M.H. Md Khir, J.O. Dennis, *et al.* A review of actuation and sensing mechanisms in MEMS-based sensor devices. *Nanoscale Research Letters*, 16:1–21, 2021.

[3] T. Almas, R. Haider, J. Malik, *et al.* Nanotechnology in interventional cardiology: A state-of-the-art review. *IJC Heart & Vasculature*, 43:101149, 2022.

[4] A.A. Babadi, S. Bagheri, and S.B.A. Hamid. Progress on implantable biofuel cell: Nano-carbon functionalization for enzyme immobilization enhancement. *Biosensors and Bioelectronics*, 79:850–860, 2016.

[5] S. Banerjee, B. Campbell, J. Rising, A. Coukell, and A. Sedrakyan. Long-term active surveillance of implantable medical devices: An analysis of factors determining whether current registries are adequate to expose safety

and efficacy problems. *BMJ Surgery, Interventions, & Health Technologies*, 1(1):e000011, 2019.

[6]   R.-D. Battmer, G.M. O'Donoghue, and T. Lenarz. A multicenter study of device failure in European Cochlear Implant Centers. *Ear and hearing*, 28(2):95S–99S, 2007.

[7]   J. Bergsland, O. J. Elle, and E. Fosse. Barriers to medical device innovation. *Medical devices (Auckland, NZ)*, 7:205, 2014.

[8]   K. Bhargava, V. Arora, A. Jaswal, *et al.* Premature battery depletion with St. Jude medical ICD and CER-D devices: Indian heart rhythm society guidelines for physicians. *Indian Heart Journal*, 71(1):12–14, 2019.

[9]   A. Björklund and U. Stenevi. Intracerebral neural implants: Neuronal replacement and reconstruction of damaged circuitries. *Annual Review of Neuroscience*, 7(1):279–308, 1984.

[10]   P.J. Blackshear, F.D. Dorman, P.L. Blackshear, R.L. Varco, and H. Buchwald. The design and initial testing of an implantable infusion pump. *Surgery Gynecology and Obstetrics*, 134(1):51–56, 1972.

[11]   G. Boopathy, K. Gurusami, K.R. Vijayakumar, and M.H. Kumar. Nanocoatings in medicine: Revolutionizing healthcare through precision and potential. In *Sustainable Approach to Protective Nanocoatings*, pages 265–289. IGI Global, 2024.

[12]   M. Bteich, J. Hanna, J. Costantine, *et al.* A non-invasive flexible glucose monitoring sensor using a broadband reject filter. *IEEE Journal of Electromagnetics, RF and Microwaves in Medicine and Biology*, 5(2):139–147, 2020.

[13]   P. Buckle, P.J. Clarkson, R. Coleman, *et al.*. Design for patient safety. *Department of Health, London*, 2003.

[14]   J. Butler, R.D. Handy, M. Upton, and A. Besinis. Review of antimicrobial nanocoatings in medicine and dentistry: Mechanisms of action, biocompatibility performance, safety, and benefits compared to antibiotics. *ACS Nano*, 17(8):7064–7092, 2023.

[15]   P. Carneiro, M.P.S. dos Santos, A. Rodrigues, *et al.* Electromagnetic energy harvesting using magnetic levitation architectures: A review. *Applied Energy*, 260:114191, 2020.

[16]   A.P. Chandrakasan, N. Verma, and D.C. Daly. Ultralow-power electronics for biomedical applications. *Annual Review of Biomedical Engineering*, 10(1):247–274, 2008.

[17]   K. Chen, J. Ren, C. Chen, W. Xu, and S. Zhang. Safety and effectiveness evaluation of flexible electronic materials for next generation wearable and implantable medical devices. *Nano Today*, 35:100939, 2020.

[18]   C. Chircov and A.M. Grumezescu. Microelectromechanical systems (MEMS) for biomedical applications. *Micromachines*, 13(2):164, 2022.

[19]   DAIC Staff. St. Jude medical recalls ICDs and CRT-D due to premature battery depletion, 2016. Accessed: June 7, 2025. Reports on device recall related to lithium cluster-induced battery failure and patient death risk.

[20]   Abbott Implantable Cardiac Devices. Battery performance alert for some. 2018. https://www.cardiovascular.abbott/content/dam/cv/cardiovascular/pdf/reports/Battery-PerformanceAlert-Whitepaper-April2018-26663-SJM-CRM-0218-0120-1.pdf

[21]   J.P. DiMarco. Implantable cardioverter–defibrillators. *New England Journal of Medicine*, 349(19):1836–1847, 2003.

[22]   EU MDR Resource Center. Post-market surveillance system under EU MDR, 2024. Accessed: June 7, 2025. Discusses Article 83 and its distinction from vigilance requirements.

[23]   A.O. Everhart, P. Karaca-Mandic, R.F. Redberg, J.S. Ross, and S.S. Dhruva. Late adverse event reporting from medical device manufacturers to the US Food and Drug Administration: Cross-sectional study. *BMJ*, 388, 2025.

[24]   Y. Fan, X. Liu, and C. Xu. A broad dual-band implantable antenna for RF energy harvesting and data transmitting. *Micromachines*, 13(4), 2022.

[25]   A. Fanelli and D. Ghezzi. Transient electronics: New opportunities for implantable neurotechnology. *Current Opinion in Biotechnology*, 72:22–28, 2021.

[26]   P. Feng, P. Yeon, Y. Cheng, M. Ghovanloo, and T.G. Constandinou. Chip-scale coils for millimeter-sized bio-implants. *IEEE Transactions on Biomedical Circuits and Systems*, 12(5):1088–1099, 2018.

[27]   M.A. Gabris and J. Ping. Carbon nanomaterial-based nanogenerators for harvesting energy from environment. *Nano Energy*, 90:106494, 2021.

[28]   A.R. Gagliardi, A. Ducey, P. Lehoux, *et al.* Factors influencing the reporting of adverse medical device events: Qualitative interviews with physicians about higher risk implantable devices. *BMJ Quality & Safety*, 27(3): 190–198, 2018.

[29]   O.Z. Gall, C. Meng, H. Bhamra, *et al.* A batteryless energy harvesting storage system for implantable medical devices demonstrated in situ. *Circuits, Systems, and Signal Processing*, 38:1360–1373, 2019.

[30]   W. Gao, Y. Wang, and F. Lai. Thermoelectric energy harvesting for personalized healthcare. *Smart Medicine*, 1(1):e20220016, 2022.

[31]   Z. Gao, Y. Zhou, J. Zhang, *et al.* Advanced energy harvesters and energy storage for powering wearable and implantable medical devices. *Advanced Materials*, 36(43):1360–1373, 2024. https://doi.org/10.1002/adma.202404492

[32]   K. Garmer, E. Liljegren, A.-L. Osvalder, and S. Dahlman. Arguing for the need of triangulation and iteration when designing medical equipment. *Journal of clinical monitoring and computing*, 17:105–114, 2002.

[33]   M. Gholikhani, S.Y. Beheshti Shirazi, G.M. Mabrouk, and S. Dessouky. Dual electromagnetic energy harvesting technology for sustainable transportation systems. *Energy Conversion and Management*, 230:113804, 2021.

[34]   S. Gibney, J.M. Hicks, A. Robinson, A. Jain, P. Sanjuan-Alberte, and F.J. Rawson. Toward nanobioelectronic medicine: Unlocking new applications using nanotechnology. *Wiley Interdisciplinary Reviews: Nanomedicine and Nanobiotechnology*, 13(3):e1693, 2021.

[35]   M. Gilasgar, A. Barlabé, and L. Pradell. High-efficiency reconfigurable dual-band class-F power amplifier with harmonic control network using MEMS. *IEEE Microwave and Wireless Components Letters*, 30(7):677–680, 2020.

[36]   A.C.R. Grayson, R.S. Shawgo, A.M. Johnson, *et al.* A BioMEMS review: MEMS technology for physiologically integrated devices. *Proceedings of the IEEE*, 92(1):6–21, 2004.

[37]   J. Gutiérrez-Martínez, C. Toledo-Peral, J. Mercado-Gutiérrez, A. Vera-Hernández, and L. Leija-Salas. Neuroprosthesis devices based on micro-and nanosensors: A systematic review. *Journal of Sensors*, 2020(1):8865889, 2020.

[38]   R.R. Harrison and C. Charles. A low-power low-noise CMOS amplifier for neural recording applications. *IEEE Journal of solid-state circuits*, 38(6):958–965, 2003.

[39]   M.A.M. Hasan, Y. Wang, C.R. Bowen, and Y. Yang. 2D nanomaterials for effective energy scavenging. *Nano-Micro Letters*, 13:1–41, 2021.

[40]   M. Hassan, G. Abbas, N. Li, *et al.* Significance of flexible substrates for wearable and implantable devices: Recent advances and perspectives. *Advanced Materials Technologies*, 7(3):2100773, 2022.

[41]   A.M. Hodge. Overview of IEEE research in biomedical circuits and systems. In *2009 IEEE/NIH Life Science Systems and Applications Workshop*, pages 17–20. IEEE, 2009.

[42]   M.-L. Hsieh, S.-K. Yeh, J.-H. Lee, M.-C. Cheng, and W. Fang. CMOS-MEMS capacitive tactile sensor with vertically integrated sensing electrode array for sensitivity enhancement. *Sensors and Actuators A: Physical*, 317:112350, 2021.

[43]   N. Jackson, L. Keeney, and A. Mathewson. Flexible-CMOS and biocompatible piezoelectric ALN material for MEMS applications. *Smart Materials and Structures*, 22(11):115033, 2013.

[44]   D. Jiang, B. Shi, H. Ouyang, Y. Fan, Z.L. Wang, and Z. Li. Emerging implantable energy harvesters and self-powered implantable medical electronics. *ACS Nano*, 14(6):6436–6448, 2020.

[45]   G. Jiang and D.D. Zhou. Technology advances and challenges in hermetic packaging for implantable medical devices. In *Implantable Neural Prostheses 2*, pages 27–61. Springer, 2009.

[46]   A. Jin, Y. Wang, K. Lin, and L. Jiang. Nanoparticles modified by polydopamine: Working as "drug" carriers. *Bioactive Materials*, 5(3):522–541, 2020.

[47]   X. Jin, G. Li, T. Xu, L. Su, D. Yan, and X. Zhang. Fully integrated flexible biosensor for wearable continuous glucose monitoring. *Biosensors and Bioelectronics*, 196:113760, 2022.

[48]   Yeun-Ho Joung. Development of implantable medical devices: From an engineering perspective. *International Neurourology Journal*, 17(3):98, 2013.

[49]   M. Karimzadehkhouei, B. Ali, M.J. Ghourichaei, and B.E. Alaca. Silicon nanowires driving miniaturization of microelectromechanical

systems physical sensors: A review. *Advanced Engineering Materials*, 25(12):2300007, 2023.

[50] H.-J. Kim, H. Hirayama, S. Kim, K.J. Han, R. Zhang, and J.-W. Choi. Review of near-field wireless power and communication for biomedical applications. *IEEE Access*, 5:21264–21285, 2017.

[51] L.W. Kleiner, J.C. Wright, and Y. Wang. Evolution of implantable and insertable drug delivery systems. *Journal of controlled release*, 181:1–10, 2014.

[52] G. Kotzar, M. Freas, P. Abel, *et al.* Evaluation of MEMS materials of construction for implantable medical devices. *Biomaterials*, 23(13):2737–2750, 2002.

[53] B.A. Kuzubasoglu and S.K. Bahadir. Flexible temperature sensors: A review. *Sensors and Actuators A: Physical*, 315:112282, 2020.

[54] X. Le, Q. Shi, P. Vachon, E.J. Ng, and C. Lee. Piezoelectric MEMS—evolution from sensing technology to diversified applications in the 5G/Internet of Things (IoT) era. *Journal of Micromechanics and Microengineering*, 32(1):014005, 2021.

[55] J.Y. Lee, H.Y. Shin, S.W. Kang, C. Park, and S.W. Kim. Improvement of electrical properties via glucose oxidase-immobilization by actively turning over glucose for an enzyme-based biofuel cell modified with DNA-wrapped single walled nanotubes. *Biosensors and Bioelectronics*, 26(5):2685–2688, 2011.

[56] M.A. Lee, S. Wang, X. Jin, *et al.* Implantable nanosensors for human steroid hormone sensing in vivo using a self-templating corona phase molecular recognition. *Advanced Healthcare Materials*, 9(21):2000429, 2020.

[57] W.H. Lee, G.D. Cha, and D.-H. Kim. Flexible and biodegradable electronic implants for diagnosis and treatment of brain diseases. *Current Opinion in Biotechnology*, 72:13–21, 2021.

[58] E. Liljegren, A.-L. Osvalder, and S. Dahlman. Setting the requirements for a user-friendly infusion pump. In *Proceedings of the Human Factors and Ergonomics Society Annual Meeting*, volume 44, pages 132–135. SAGE Publications Sage CA: Los Angeles, CA, 2000.

[59] K.D. Lind. Understanding the market for implantable medical devices. *Insight*, 2017.

[60] N.R. Lindquist, N.D. Cass, A. Patro, *et al.* Hires ultra series recall: Failure rates and revision speech recognition outcomes. *Otology & Neurotology*, 43(7):e738–e745, 2022.

[61] S. Liu, Y. Zhao, W. Hao, X.-D. Zhang, and D. Ming. Micro-and nanotechnology for neural electrode-tissue interfaces. *Biosensors and Bioelectronics*, 170:112645, 2020.

[62] Y. Luo, M.R. Abidian, J.-H. Ahn, *et al.* Technology roadmap for flexible sensors. *ACS Nano*, 17(6):5211–5295, 2023.

[63] D. Maddipatla, B.B. Narakathu, and M. Atashbar. Recent progress in manufacturing techniques of printed and flexible sensors: A review. *Biosensors*, 10(12):199, 2020.

[64]   J.L. Martin, B.J. Norris, E. Murphy, and J.A. Crowe. Medical device development: The challenge for ergonomics. *Applied Ergonomics*, 39(3): 271–283, 2008.

[65]   E. Meng and R. Sheybani. Insight: Implantable medical devices. *Lab on a Chip*, 14(17):3233–3240, 2014.

[66]   R.T. Miyamoto, M. Young, W.A. Myres, K. Kessler, K. Wolfert, and K.I. Kirk. Complications of pediatric cochlear implantation. *European Archives of Oto-Rhino-Laryngology*, 253:1–4, 1996.

[67]   J.A. Montgomery, J.Y. Sensing, S.D. Saunders, *et al.* Premature battery depletion due to compromised low-voltage capacitor in a family of defibrillators. *Pacing and Clinical Electrophysiology*, 42(7):965–969, 2019.

[68]   T.M. Morrison, P. Pathmanathan, M. Adwan, and E. Margerrison. Advancing regulatory science with computational modeling for medical devices at the FDA's Office of Science and Engineering Laboratories. *Frontiers in Medicine*, 5, 2018.

[69]   R. Narayanamoorthi. Modeling of capacitive resonant wireless power and data transfer to deep biomedical implants. *IEEE Transactions on Components, Packaging and Manufacturing Technology*, 9(7):1253–1263, 2019.

[70]   S. Panda, S. Hajra, K. Mistewicz, *et al.* Piezoelectric energy harvesting systems for biomedical applications. *Nano Energy*, 100:107514, 2022.

[71]   J. Pane, K.M.C. Verhamme, L. Shrum, I. Rebollo, and M.C.J.M. Sturkenboom. Blockchain technology applications to postmarket surveillance of medical devices. *Expert Review of Medical Devices*, 17(10):1123–1132, 2020.

[72]   P. Pattanaik and M. Ojha. Review on challenges in MEMS technology. *Materials Today: Proceedings*, 81:224–226, 2023.

[73]   J.J. Percy and S. Kanthamani. Revolutionizing wireless communication: A review perspective on design and optimization of RF MEMS switches. *Microelectronics Journal*, 139:105891, 2023.

[74]   R.M.R. Pinto, V. Gund, R. A. Dias, K.K. Nagaraja, and K.B. Vinayakumar. CMOS-integrated aluminum nitride MEMS: A review. *Journal of Microelectromechanical Systems*, 31(4):500–523, 2022.

[75]   Y. Polo, J. Luzuriaga, J. Iturri, *et al.* Nanostructured scaffolds based on bioresorbable polymers and graphene oxide induce the aligned migration and accelerate the neuronal differentiation of neural stem cells. *Nanomedicine: Nanotechnology, Biology and Medicine*, 31:102314, 2021.

[76]   G.R. Prasad, B.T.P. Madhav, P. Pardhasaradhi, *et al.* Concentric ring structured reconfigurable antenna using MEMS switches for wireless communication applications. *Wireless Personal Communications*, 120:587–608, 2021.

[77]   N. Quack, A.Y. Takabayashi, H. Sattari, *et al.* Integrated silicon photonic MEMS. *Microsystems & Nanoengineering*, 9(1):27, 2023.

[78]   A. Rashidi, N. Yazdani, and A.M. Sodagar. Fully-integrated, high-efficiency, multi-output charge pump for high-density microstimulators. In *2018 IEEE Life Sciences Conference (LSC)*, pages 291–294. IEEE, 2018.

[79]   U. Rawat, J.D. Anderson, and D. Weinstein. Design and applications of integrated transducers in commercial CMOS technology. *Frontiers in Mechanical Engineering*, 8:902421, 2022.

[80]   M.R. Sarker, M.H. Md Saad, J.L. Olazagoitia, and J. Vinolas. Review of power converter impact of electromagnetic energy harvesting circuits and devices for autonomous sensor applications. *Electronics*, 10(9):1108, 2021.

[81]   L. Shabani, M. Abbasi, Z. Azarnew, A.M. Amani, and A. Vaez. Neuro-nanotechnology: Diagnostic and therapeutic nano-based strategies in applied neuroscience. *Biomedical Engineering Online*, 22(1):1, 2023.

[82]   D.T. Shakir, H.J. Al-Qureshy, and S.S. Hreshee. Performance analysis of MEMS-based oscillator for high frequency wireless communication systems. *International Journal of Communication Networks and Information Security*, 14(3):86–98, 2022.

[83]   B. Shi, Z. Li, and Y. Fan. Implantable energy-harvesting devices. *Advanced Materials*, 30(44):1801511, 2018.

[84]   Y. Shi, R. Liu, L. He, H. Feng, Y. Li, and Z. Li. Recent development of implantable and flexible nerve electrodes. *Smart Materials in Medicine*, 1:131–147, 2020.

[85]   M. Shikida, Y. Hasegawa, M.S. Al Farisi, M. Matsushima, and T. Kawabe. Advancements in MEMS technology for medical applications: Microneedles and miniaturized sensors. *Japanese Journal of Applied Physics*, 61(SA):SA0803, 2021.

[86]   M.M.H. Shuvo, T. Titirsha, N. Amin, and S.K. Islam. Energy harvesting in implantable and wearable medical devices for enduring precision healthcare. *Energies*, 15(20):7495, 2022.

[87]   R. Tarricone, O. Ciani, A. Torbica, *et al.* Lifecycle evidence requirements for high-risk implantable medical devices: A European perspective. *Expert Review of Medical Devices*, 17(10):993–1006, 2020.

[88]   M. Tilli, M. Paulasto-Kröckel, M. Petzold, H. Theuss, T. Motooka, and V. Lindroos. *Handbook of silicon-based MEMS materials and technologies.* Elsevier, 2020.

[89]   F.V.Y. Tjong and V.Y. Reddy. Permanent leadless cardiac pacemaker therapy: A comprehensive review. *Circulation*, 135(15):1458–1470, 2017.

[90]   M. Toledano, J. Gutierrez-Pérez, A. Gutierrez-Corrales, *et al.* Novel non-resorbable polymeric-nanostructured scaffolds for guided bone regeneration. *Clinical Oral Investigations*, 24:2037–2049, 2020.

[91]   B. Upendra, B. Panigrahi, K. Singh, and G.R. Sabareesh. Recent advancements in piezoelectric energy harvesting for implantable medical devices. *Journal of Intelligent Material Systems and Structures*, 35(2):129–155, 2024.

[92]   U.S. Food and Drug Administration. Strengthening our national system for medical device postmarket surveillance. White paper, Center for Devices and Radiological Health, FDA, 2013. Accessed: June 7, 2025. Discusses short life cycle challenges compared to drugs and biologics.

[93]   U.S. Government Accountability Office. Real world evidence: A roundtable to identify opportunities, challenges, and solutions in medical device postmarket surveillance. Technical Report GAO-24-106699, U.S. Government Accountability Office, 2024. Accessed: June 7, 2025.

[94]   M. Viceconti, F. Pappalardo, B. Rodriguez, M. Horner, J. Bischoff, and F.M. Tshinanu. In silico trials: Verification, validation and uncertainty quantification of predictive models used in the regulatory evaluation of biomedical products. *Methods*, 185:120–127, 2021.

[95]   G. Voskerician, M.S. Shive, R.S. Shawgo, *et al.* Biocompatibility and biofouling of MEMS drug delivery devices. *Biomaterials*, 24(11):1959–1967, 2003.

[96]   J. Wang, T. Wang, H. Liu, *et al.* Flexible electrodes for brain–computer interface system. *Advanced Materials*, 35(47):2211012, 2023.

[97]   L. Wang, K. Jiang, and G. Shen. Wearable, implantable, and interventional medical devices based on smart electronic skins. *Advanced Materials Technologies*, 6(6):2100107, 2021.

[98]   J.R. Ward and P.J. Clarkson. An analysis of medical device-related errors: Prevalence and possible solutions. *Journal of Medical Engineering & Technology*, 28(1):2–21, 2004.

[99]   J.D. Weiland and M.S. Humayun. Retinal prosthesis. *IEEE Transactions on Biomedical Engineering*, 61(5):1412–1424, 2014.

[100]   L.S.Y. Wong, S. Hossain, A. Ta, J. Edvinsson, D.H. Rivas, and H. Naas. A very low-power CMOS mixed-signal IC for implantable pacemaker applications. *IEEE Journal of Solid-State Circuits*, 39(12):2446–2456, 2004.

[101]   C.-H. Xu, P.-J. Ye, Y.-C. Zhou, D.-X. He, H. Wei, and C.-Y. Yu. Cell membrane-camouflaged nanoparticles as drug carriers for cancer therapy. *Acta Biomaterialia*, 105:1–14, 2020.

[102]   Z. Xu, C. Jin, A. Cabe, *et al.* Flexible energy harvester on a pacemaker lead using multibeam piezoelectric composite thin films. *ACS Applied Materials & Interfaces*, 12(30):34170–34179, 2020.

[103]   C. Yang, B. Hu, L. Lu, Z. Wang, W. Liu, and C. Sun. A miniaturized piezoelectric MEMS accelerometer with polygon topological cantilever structure. *Micromachines*, 13(10):1608, 2022.

[104]   J. Yu, W. Ling, Y. Li, *et al.* A multichannel flexible optoelectronic fiber device for distributed implantable neurological stimulation and monitoring. *Small*, 17(4):2005925, 2021.

[105]   F.-G. Zeng, S. Rebscher, W. Harrison, X. Sun, and H. Feng. Cochlear implants: System design, integration, and evaluation. *IEEE Reviews in Biomedical Engineering*, 1:115–142, 2008.

[106]   Q. Zhang, G. Zhao, Z. Li, *et al.* Multi-functional adhesive hydrogel as biointerface for wireless transient pacemaker. *Biosensors and Bioelectronics*, 263:116597, 2024.

[107]   Y. Zhang, L. Zhou, C. Liu, *et al.* Self-powered pacemaker based on all-in-one flexible piezoelectric nanogenerator. *Nano Energy*, 99:107420, 2022.

[108] H. Zhao, R. Liu, H. Zhang, P. Cao, Z. Liu, and Y. Li. Research progress on the flexibility of an implantable neural microelectrode. *Micromachines*, 13(3):386, 2022.

[109] J. Zhao, R. Ghannam, K.O. Htet, *et al.* Self-powered implantable medical devices: Photovoltaic energy harvesting review. *Advanced Healthcare Materials*, 9(17):2000779, 2020.

[110] Z. Zhao, Y. Kong, X. Lin, *et al.* Oxide nanomembrane induced assembly of a functional smart fiber composite with nanoporosity for an ultrasensitive flexible glucose sensor. *Journal of Materials Chemistry A*, 8(48): 26119–26129, 2020.

[111] M. Zhou, L. Deng, D. Wen, L. Shang, L. Jin, and S. Dong. Highly ordered mesoporous carbons-based glucose/O2 biofuel cell. *Biosensors and Bioelectronics*, 24(9):2904–2908, 2009.

[112] L. Zhu, D. Luo, and Y. Liu. Effect of the nano/microscale structure of biomaterial scaffolds on bone regeneration. *International Journal of Oral Science*, 12(1):6, 2020.

Chapter 2
Energy harvesting in biomedical implantable device

The rapid advancement of biomedical technology has driven an increasing demand for implantable medical devices capable of monitoring, diagnosing, and treating a variety of medical conditions. Although conventional battery-powered devices have proven effective in these roles, they possess limitations such as limited lifespan, the necessity for replacement procedures, and dependence on large, inflexible power sources. To overcome these challenges, researchers have developed energy harvesting methods that aim to directly capture energy from the human body or its environment, thus providing medical implants with a sustainable and potentially unlimited power source. This chapter examines the current state and potential of energy harvesting technologies in biomedical implants, exploring both established and emerging approaches to power generation and storage. Section 2.1 reviews lithium-ion batteries and other traditional energy storage systems frequently employed in medical implants, highlighting their limitations in terms of size, lifespan, and compatibility with the human body. An understanding of these constraints provides context for the transition to alternative energy options. Section 2.2 addresses state-of-the-art emerging implanted energy harvesting methodologies, including radio frequency (RF), thermal, kinetic, solar, and biochemical energy harvesting. These techniques employ body heat, movement, light exposure, biological reactions, or ambient RF signals to generate energy. The underlying theories of each technology are examined, with attention to their unique advantages and challenges as they apply to the human body. Section 2.3 assesses the performance, viability, and adaptability of implantable energy supply systems across various biomedical applications, considering factors such as practicality, biocompatibility, efficiency, and energy output. By comparing this technology, the section indicates which energy sources are suitable for different types of medical implants, ranging from complex functional neurostimulators to low-power pacemakers. Finally, Section 2.4 explores the evolution of implantable energy harvesting technologies, emphasizing recent advancements, materials, and hybrid systems that integrate multiple energy sources. This section also discusses current research and potential future developments, such as the prospect of fully autonomous, long-lasting implants, and the integration of energy harvesting technologies with intelligent power management systems. Overall, this chapter provides a comprehensive analysis of energy harvesting in biomedical implantable devices, offering insights into how these technologies can enhance the usability, safety, and comfort of future medical implants.

2.1   Conventional energy storage device

For many decades, traditional energy storage systems, such as batteries and capacitors, have provided power for biomedical and implantable devices. These technologies are crucial for operating essential medical equipment, like pacemakers, cochlear implants, neurostimulators, and drug delivery systems, all of which rely on tailored power sources to work effectively within the human body. The evolution of energy storage has paralleled implantable medical devices, focusing on longevity, miniaturization, and safety in harmony with biological tissues. Battery technology for implantable biomedical devices has consistently evolved alongside advancements in power density, lifespan, and safety. In the mid-twentieth century, mercury–zinc batteries were considered advanced technology, leading to the development of the first implanted devices. However, these cumbersome and short-lived batteries necessitated frequent replacement, which was impractical for long-term use due to the need for invasive procedures during every battery change. The advent of lithium–iodine batteries in the 1970s marked a pivotal shift with their higher energy density, stability, and extended life, setting a new standard for medical implants. Unlike their mercury-based predecessors, lithium–iodine batteries are significantly more appropriate for long-term, low-power usages, capable of powering devices such as pacemakers for years before requiring replacement. Lithium and iodine chemistry continues to be pivotal for systems demanding stability, especially in low-power settings. As lithium-based technologies advanced, the introduction of lithium-ion and lithium-polymer batteries further enriched the field. These batteries offer higher energy densities and rechargeability, a major leap forward for implantable devices. They greatly reduce the necessity for surgical replacements and extend the functional life of the devices by permitting rechargeable designs. The development of smaller, more discreet implantable devices has been facilitated by the increased energy density and reduced size of lithium-ion cells, enhancing patient comfort and expanding potential applications of medical implants. While lithium-ion technology brings significant benefits, safety concerns must also be addressed. Lithium-ion batteries can overheat, leak hazardous chemicals, and pose challenges in ensuring compatibility with human tissues. Due to these safety challenges, there has been an impetus to research materials and encapsulation techniques that mitigate risks, aiming to make batteries safer for use in sensitive and biologically active environments.

2.1.1   *Rechargeable battery system in biomedical implantable application*

Recent advancements in medical implants include enhanced capabilities driven by rechargeable battery systems such as transcutaneous inductive charging. This technology allows energy transfer from an external source to the implant wirelessly, eliminating the need for invasive procedures. This is particularly advantageous for devices requiring regular power, like neurostimulators used for pain management. Despite these benefits, rechargeable batteries present challenges as well, including degradation over time and the inconvenience of needing frequent recharges. Additionally, the lifespan of these batteries is constrained by charge–discharge cycles, necessitating further investigation into materials that can endure numerous cycles with minimal deterioration.

2.1.2 Development of capacitors in implantable devices

Battery technology and capacitors have evolved together, playing pivotal roles in powering biomedical implants. For devices such as defibrillators that require sudden, high-power outputs, capacitors provide a notable benefit due to their capacity for swift energy release. Initially, capacitors were not designed to be small or biocompatible, restricting their use in implants. Over time, advances in materials science have led to the creation of smaller and more durable electrolytic and ceramic capacitors that are ideal for implantable devices due to their compact size. More recently, supercapacitors and ultracapacitors have been developed, which can store greater energy amounts than traditional capacitors, offering rapid energy discharge when needed. These high-capacity capacitors are beneficial in hybrid power systems as they manage energy surges, reduce the load on batteries, and extend their lifespan. Capacitors also offer safety and durability benefits over batteries, as they store energy electrostatically rather than chemically, reducing risks such as leaks, overheating, and degradation over time. This makes capacitors a suitable choice for applications, like implants, where stability and long-term reliability are crucial.

2.1.3 Biocompatibility and miniaturization for conventional energy storage

Battery and capacitor technologies have evolved alongside the complexity of implantable devices, fueled by the need for smaller and more biocompatible energy storage solutions. Earlier implantable batteries were large and required robust protective casings to prevent leaks and ensure safety. With improvements in encapsulation techniques and biocompatible coatings, both capacitors and batteries can now be safely used within the human body without adversely affecting nearby tissues. These coatings, made from titanium and specialized polymers, offer longer device lifespans and act as effective barriers against chemical leaks. Furthermore, there has been a shift toward miniaturization in battery and capacitor technology. The creation of thin-film and microbatteries has paved the way for ultra-compact, lightweight energy storage systems compatible with minimally invasive implants. These technologies are particularly suited for microstimulators, biosensors, and other small medical devices where patient comfort and limited space are critical. Additionally, the innovation of nanoscale capacitors, which accommodate the compact dimensions required for modern biomedical devices, has enabled the production of capacitors with high power densities while occupying minimal physical space.

2.1.4   *The limitations and future development of conventional energy storage*

Despite recent advances, traditional batteries and capacitors still have drawbacks that could adversely affect the future of implantable device development. While recent improvements have enhanced rechargeability, the short lifespan of conventional batteries remains a significant obstacle for achieving truly long-term, self-sufficient implants. Capacitors, while ideal for applications requiring rapid, high-power output, lack the energy density needed for extended, continuous operation in implants. Ongoing research targets overcoming these limitations by improving energy density, durability, and biocompatibility for both capacitors and batteries. Notable material innovations include solid-state electrolytes and nanoengineered electrodes, which enhance cycle life and charge capacity in batteries. In capacitors, advancements in nanotechnology could boost power density and narrow the gap between the energy storage of batteries and the discharge capabilities of capacitors, paving the way for hybrid energy systems. These systems combine batteries and capacitors to provide steady power and quick energy release. Moreover, integrating batteries and capacitors with energy harvesting methods – such as thermal, kinetic, or biochemical – offers a promising approach to developing more sustainable implant power sources. Hybrid systems capable of deriving energy from the environment or body can improve patient quality of life by reducing the need for battery replacements and extending the lifespan of devices. Advances in battery and capacitor technologies have significantly bolstered the development of biomedical implants, enabling critical medical instruments to possess reliable, compact, and increasingly safe power sources. From earlier mercury–zinc batteries to modern lithium-ion cells and advanced capacitors, these innovations have extended implant lifespans and improved patient outcomes. Nevertheless, challenges related to longevity, biocompatibility, and energy density continue to drive innovation in the field. It is expected that future material science developments, miniaturization, and hybrid systems will markedly boost the performance of batteries and capacitors as research progresses. As this field advances, we move closer to an era of energy-autonomous, minimally invasive medical devices that enhance clinical outcomes, patient comfort, and reliability.

2.2   Emerging implantable energy harvesting technology

Implantable technologies that harvest energy represent a revolutionary shift in powering biomedical devices. These cutting-edge solutions extract energy from the body or nearby environment, offering an endless and self-reliant power source, unlike traditional batteries, which have finite lifespans and require frequent replacements. Techniques such as biochemical, thermal, kinetic, photovoltaic (PV), and RF energy capture convert environmental or bodily energy into electrical power, which can be stored in device batteries or capacitors. This transition from conventional energy storage methods to sustainable and independent energy acquisition paves the way for more durable, low-maintenance implants that enhance patient comfort and quality of life.

Thermal energy harvesting technology: Thermal energy harvesting involves utilizing devices like thermoelectric generators that can transform the natural heat emitted by the human body into functional electrical energy. These generators work by exploiting the temperature contrast between the human body and the ambient environment to create a voltage. This mechanism effectively converts bodily heat into electricity. Given the human body's ability to produce a relatively stable and constant heat source, this approach is particularly promising for use in implanted devices. The primary challenge lies in achieving efficient energy conversion with the body's low-temperature gradients. Recent progress in thermoelectric materials, such as the development of nanostructured materials and bismuth telluride, has enhanced the efficiency of thermoelectric generators. This advancement allows them to generate more power even at smaller temperature differences, a critical aspect for their use in biomedical implants.

Kinetic energy harvesting technology: Also known as mechanical energy harvesting, this process involves harnessing kinetic energy generated by bodily movements, blood circulation, joint movements, and other muscle actions. Devices like triboelectric, piezoelectric, and electromagnetic systems are capable of capturing this energy form. To record continuous movements, piezoelectric materials may be positioned near active organs, such as the heart or lungs, as they produce an electrical charge when subjected to mechanical stress. The operation of triboelectric nanogenerators (TENGs) relies on the friction between two energy-harvesting materials, whereas electromagnetic energy harvesters utilize magnetic induction to produce electricity. Despite its significant potential, kinetic energy harvesting is frequently challenged by the unpredictable nature of body movements, necessitating effective storage solutions to ensure a stable power supply.

PV energy harvesting technology: PV energy harvesting, which involves using light to produce electrical power, is increasingly being explored for implantable devices situated near the skin or in areas exposed to light. Subcutaneous PV cells harness transdermal illumination or environmental light to generate electricity. Advancements in PV technology, like the development of flexible and biocompatible organic PV cells, have expanded the potential for these cells in biomedical devices. These cells conform to the body's shape to optimize light absorption and provide power for implants such as biosensors. However, the main challenge remains achieving sufficient light penetration and energy production, particularly in low-light or dark environments.

Biochemical energy harvesting: In biochemical energy harvesting, or biofuel cell (BC) technology, the body's biochemical activities are harnessed as an energy source. By mimicking the body's metabolic pathways, these systems, often using fuel cells based on glucose or lactate, generate electricity. Glucose-based BCs, for example, utilize the body's glucose and oxygen to produce a steady, low-power current, making them ideal for devices like medical delivery systems or glucose monitors. As BCs rely on naturally occurring substances in the body, they are inherently biocompatible. However, due to their currently low-power output, significant research efforts are focused on improving their lifespan and efficiency to maintain proper function over extended durations without degradation.

RF energy harvesting technology: Sources of electromagnetic energy external to the implant that can be used for RF energy harvesting include Wi-Fi signals, cell networks, and specific RF transmitters. This makes RF energy harvesting highly beneficial because the implant can consistently receive power as long as it remains within the RF source's range. RF energy's dual ability to facilitate wireless data transfer and power supply makes it particularly suitable for implants with communication features. However, there are notable challenges due to the limited penetration of RF signals and the attenuation caused by body tissues. Extensive studies on resonance-matching techniques and antenna optimization are underway to improve the body's ability to capture RF energy.

Hybrid energy harvesting systems: Due to the distinct nature of each energy harvesting method, hybrid systems have become a viable option for maintaining a consistent and dependable power supply in implants. Hybrid energy harvesting incorporates multiple energy sources such as thermal, kinetic, biological, and RF energies to capitalize on the benefits of all. This synergy not only boosts the flexibility of implants to adjust to changing environmental conditions or shifts in energy supply but also ensures a stable power output. For example, a hybrid system might harness kinetic energy during active body movements and switch to thermal or RF energy during inactivity. This strategy keeps power levels stable while prolonging the device's operational lifespan.

Energy harvesting systems require advancements in energy storage solutions because the energy they capture is often unstable and needs efficient storage for reliable device operation. Traditional batteries and capacitors, though still widely used, are being updated to function more effectively in hybrid systems. To mitigate power fluctuations, there's an increasing trend to pair energy harvesters with supercapacitors and microbatteries capable of absorbing and storing intermittent energy bursts. Additionally, due to their improved biocompatibility and compactness, solid-state and thin-film batteries are becoming preferred choices, fitting well with the needs of minimally invasive implants. These storage solutions ensure the device's continuous operation despite inconsistent energy collection. Nonetheless, despite the advantages that contemporary energy harvesting technologies bring to implantable devices, significant challenges remain. Biocompatibility continues to be a critical concern, as materials in energy harvesters must interact safely with human tissue without degrading inside the body. The longevity of these implants hinges on addressing overheating, leakage, and immune responses. Safety is another important issue to tackle. The efficiency of energy conversion often falls short of desired levels, necessitating ongoing research to enhance material performance for energy harvesting and optimize device design. Success in overcoming these challenges through advancements in materials science, microengineering, and power management will set the path for future implantable energy harvesting technology. Researchers are probing new materials such as graphene, nanocomposites, and bio-inspired polymers to boost the effectiveness and biocompatibility of energy harvesters. In addition, power management systems are integrating artificial intelligence (AI) and machine learning, allowing implants to autonomously adjust operations as per energy availability. New implantable energy harvesting technologies are revolutionizing biomedical devices by providing sustainable replacements for conventional batteries, achieving self-sufficient, long-lasting implants. By harnessing energy from the environment or the body, these technologies enhance patient comfort and device dependability while negating the need for invasive procedures. In the near future, the development of hybrid systems, advanced materials, and intelligent power management will lead implants to be less invasive yet highly adaptable and energy efficient. Energy harvesting could soon become an intrinsic feature of biomedical implants, with ongoing research and development pointing toward a revolution in patient care through self-sustaining and sustainable medical devices.

2.3   Comparison of the implantable energy supply technologies

Figure 2.1(a)–(j) demonstrates the power consumption of various implantable devices and the harvestable amount of power by different technologies. The voltage requirements for a commercial implantable device is 2–3 V [42]. Batteries can satisfy the power and voltage requirements for implantable devices. However, their large size and weight restrict their use. In addition, there are chemical leakage risks that can cause damage to the body [11]. Thus, for an implantable power harvester to be effective, it must have a smaller size, better lifetime and provide sustainable power.

Figure 2.2 shows a comparison between the power density that can be generated from different power harvesting approaches. Clearly, implantable PV cells can harvest the highest power density. RF energy harvesters can generate high power in the near-field, but careful alignment is necessary. Far-field RF harvesters are inefficient when the frequency is increased to tens of gigahertz. The start functioning and magnetic field mismatch are the main constraints [10,32,49]. To be specific, the RF system requires power to wake up the system because the complementary metal–oxide–semiconductor (CMOS) components always have a threshold, but far-field RF harvesters are inefficient. A perfect match between transmitting and receiving coils can increase the magnetic flux density, hence improve the power output. Large coil size, short power range, and tissue losses are lingering challenges [2,39].

As mentioned above, piezoelectric generator (PEG) and TENG can harvest kinetic energy from human motion [23]. As can be seen from Figure 2.2, the size of PEGs is larger than that of PV cells, BCs, and thermal electric generators (TEGs). Some PEGs have a small size, but the harvested amount of power is low. Another problem of PEGs and TENGs is the unstable and low output voltage. Both these technologies need to be implanted into moving organs such as the heart, lungs, or in the pericardial region. However, the size, rate, and external force of the organ affect the output voltage of the PEGs [42].

BCs are highly biocompatible with living organs, since they are often tested with plants, insects, and even mammals [18,22,30,40,45]. According to Figure 2.2, the

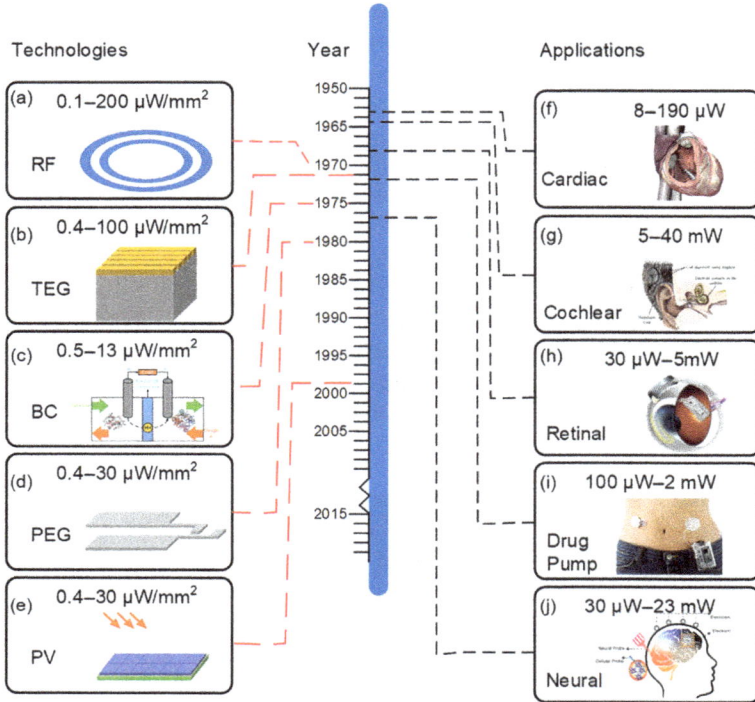

Figure 2.1   The time development of implantable power harvesters and implantable applications associated with their power range

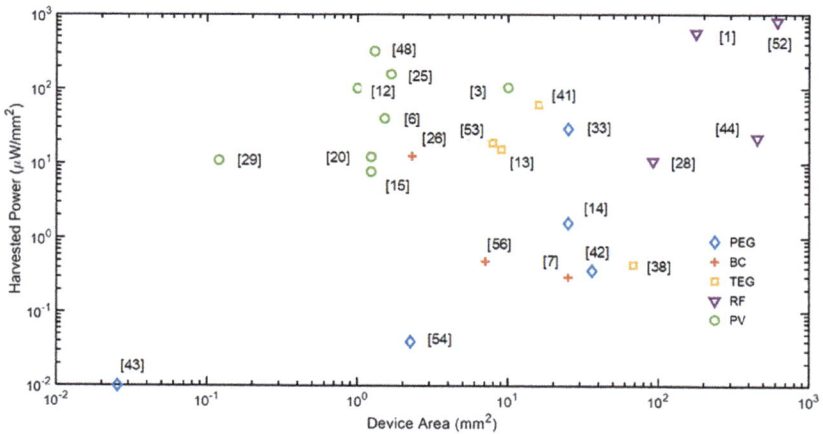

Figure 2.2   The harvested power changes of the implantable PV cells from 2012 to now: photovoltaic cell [3,6,12,15,20,25,29,48], biofuel cell [7,26,55], piezoelectric generator [14,33,42,43,53], thermal electric generator [13,38,41,52], and radio frequency [1,28,44,51]

amount of harvestable power from BCs is similar to PEGs, but their size is relatively small. Although BCs are advantageous due to cheap fabrication costs and high biocompatibility, the degradation of the enzyme is still a challenge. Compared with other power techniques, TEGs highly depend on the temperature changes, but the maximum temperature difference in the human body is less than 6 °C [8].

Implantable PV cells have commonly been used as small-scale harvesters and have been demonstrated to generate power from ambient light. In addition, the introduction of light management techniques, power management circuits, maximum power point tracking (MPPT) control logic, and startup circuits will tremendously improve the performance of implantable solar cells [6,12,19]. Most importantly, new developments in solar cell technologies have enabled multi-crystalline PV cells to achieve an efficiency improvement from 21.9% to 22.3% within one year (2017–18). Furthermore, new and emerging materials such as perovskite have enabled a 1.2% increase in solar cell efficiency.

2.4 Development of implantable energy harvesting technology

Thanks to advancements in the microelectronics industry, biomedical implantable devices [16] now play an increasingly significant role in the diagnoses, treatment, and monitoring of various diseases using miniaturized and high-resolution biosensors [27,31], reliable power transducers [54], and efficient integrated circuitry [19,35]. A variety of subcutaneous devices such as defibrillators [17], pacemakers [36], cochlear implants [5,34], drug pumps [24,37,46], as well as muscle, retinal, and neurological stimulators [47,48,50] are now being used for clinical applications. For instance, to avoid sudden heart failure, patients who suffer from heart diseases require long-term heart rhythm monitoring and analysis. Moreover, a defibrillating shock is administered during cardiac arrest. Such technologies work in harmony with personalized medical devices (e.g., wearables) and have great potential for real-world use. However, power harvesting or generation is still a main challenge in such implantable devices. Most implantable devices are powered using toxic batteries, which need eventual surgical replacement [21]. Some of the drawbacks of these batteries include their size, lifespan, and the risk of leaking toxic hazardous substances. Therefore, there is an urgent need for transducers that can harvest power from the human body or from the ambient environment. There are a variety of approaches for harvesting energy from the subcutaneous environment using PV cells, RF harvesters, PEG, TEGs, BCs, as well as other hybrid energy harvesting techniques. Figure 2.3(a) shows the RF Energy Harvesting technique that uses a retinal stimulating system [2]. Unfortunately, many of these techniques have limitations due to their large size, low output power density, or unstable energy output.

Figure 2.3 illustrates how each energy harvesting technology's power density differs greatly. With its wide power range (0.1–200 $\mu W/mm^2$), RF energy harvesting is appropriate for low-power applications such as cardiac monitoring, where device function can be supported by intermittent energy from external sources. Due to its reliance on external transmitters, RF may not be as reliable in situations requiring constant power. TEGs use the body's temperature gradient to create power, giving them a power density of 0.4–100 $\mu W/mm^2$. Deeply implanted devices, such as cochlear implants, where a constant temperature differential is required to maintain low-power

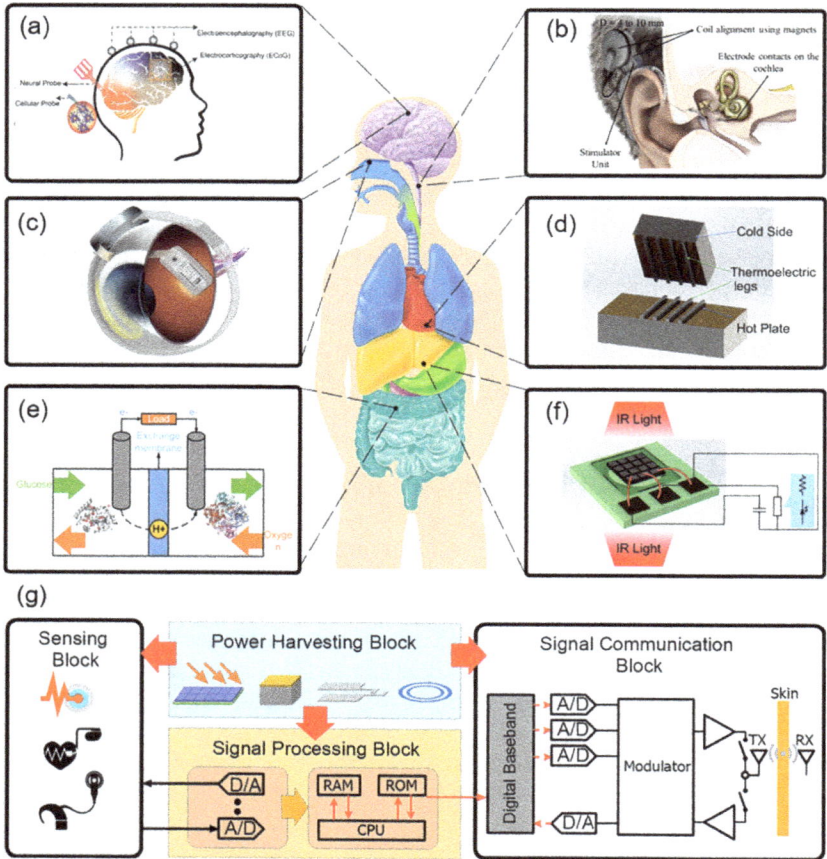

Figure 2.3   *Implantable biomedical power harvesting applications. (a) A bio-implantable system of a neurostimulator [16], (b) a piezoelectric generator in the usage of a cochlear hearing aid [34], (c) retinal application powered by wireless power transfer [4], (d) thermal electric energy harvester in application of cardiac implants [4,8], (e) an enzyme biofuel generator collecting power from the surrounding glucose, (f) PV cell in application of neurostimulator [48], and (g) the schematic diagram of the implantable applications contains four blocks: Sensing block, Power harvesting and conditioning block, Signal processing block, and the Signal communication block [9].*

operations, benefit greatly from this technique. The energy output of TEGs is limited by their low efficiency at tiny temperature gradients, which means that only devices with low power requirements may use them.

BCs (0.5–13 $\mu W/mm^2$) are becoming more popular for long-term implants because they can generate electricity from the body's biological energy, such as glucose. Applications such as retinal and brain implants, where consistent, low-power consumption is necessary for therapeutic or sensory activities, show promise for these cells. Their wider application in higher power requirements is, however, limited by their relatively low power density and the difficulties in maintaining reliable fuel sources.

PV cells and PEG power applications requiring mechanical or light energy sources are in need of power densities ranging from 0.4 to 30 $\mu W/mm^2$. Applications with mechanical motion around them – such as cardiac devices that can stimulate contractions of muscles or heartbeat – have PEGs better suited for them. The PV cells capture ambient light in shallow implants near the skin, similar to biosensors. These solutions may require additional sources of energy or storage devices to provide a constant power supply since PV cells rely on external light, and the body movements associated with PEGs are not consistent.

Improvements in energy harvesting technologies have widened the range of biomedical applications, as can also be seen from Figure 2.3. Today, various energy harvesting systems are applied to provide power for devices such as cardiac monitors, cochlear implants, retinal stimulators, medication pumps, and neurological stimulators during long-term operation. Conventionally, for instance, the cochlear implant and other cardiac devices are powered by batteries and should be replaced frequently. Devices with integrated energy harvesters could, in theory, operate indefinitely with no need for numerous surgical replacement procedures. Particularly for illnesses that need intermittent or continuous monitoring and treatment, these improvements have significant medical relevance. RF and TEG systems can provide continuous power for cardiac patients' gadgets that can monitor their heart rhythm in real time. With the use of biochemical or PEG systems, retinal and neurological devices that aid in sensory restoration and therapeutic neural stimulation may now operate independently, guaranteeing continuous operation without the need for regular battery replacements.

The promise of implantable energy-collecting devices is still far from being completely realized, despite the encouraging applications. Because small devices cannot incorporate large or complicated harvesters without sacrificing comfort and biocompatibility, the quantity of energy that may be harvested is limited by size. The integration of energy harvesters with smaller implants is limited because they frequently sacrifice size or flexibility in order to achieve higher power densities. Material compatibility is also essential; the devices must function well while being safe and biocompatible to reduce immunological reactions and avoid negative effects over time. The body's ability to convert energy efficiently and steadily is also very important. Certain environmental conditions are necessary for any energy harvesting technique to work at its best. For example, PEGs must have regular body motions, while BCs must have a steady supply of fuel. Because of this, scientists are investigating hybrid systems that integrate two or more harvesting technologies in order to increase dependability and consistency. For example, a hybrid implant may take power from both body heat and external signals (RF and TEG) at the same time, resulting in a more constant output even under a variety of environmental conditions.

The next generation of implantable energy harvesting devices is anticipated to heavily rely on developments in power management systems. AI and machine learning algorithms can be integrated into power management systems to optimize consumption according to the operational requirements of the device and automatically adjust to the availability of energy sources. AI-powered algorithms might, for example, anticipate the most effective energy sources based on patient activity levels, alternating between RF and PEG systems in response to motion or the availability of external signals. In fluctuating physiological contexts, this flexibility will be essential to guaranteeing that implants continue to be responsive and functioning. Furthermore, the potential of these technologies is probably going to be further enhanced by miniaturization and nanotechnology. In addition to enabling smaller and more flexible designs, high surface area nanomaterials like graphene can increase the energy conversion efficiency of BCs and TEGs. The integration of energy solutions into microimplants will require the development of biocompatible, nanoscale energy harvesters, creating new opportunities for minimally invasive devices that function independently inside the body.

Implantable energy harvesting technologies have reached such a revolutionary stage that in the near future, they may provide biomedical devices with longer lifespans using noninvasive implants that can help improve patient health without frequently changing batteries or undergoing surgery. At this point, all existing techniques of energy harvesting – be it RF, TEG, PEG, BC, and PV – have drawbacks, for which continuous research and development are getting over their adverse impacts. Advanced power management systems, along with a range of harvesting techniques, might well enable future implants to achieve unparalleled dependability and autonomy. We are one step closer to fully self-sustaining biomedical implants that work together in harmony with the human body and will really revolutionize patient care and quality of life. It goes without saying that the main work on the development of these devices will be done by the integration of AI, hybrid systems, and nanotechnology.

References

[1]   A. Afsahi, B. Afshar, E. Afshari, *et al.* 2010 Index IEEE Journal of Solid-State Circuits Vol. 45. *IEEE Journal of Solid-State Circuits*, 45(12):2883, 2010.

[2]   K. Agarwal, R. Jegadeesan, Y.-X. Guo, and N.V. Thakor. Wireless power transfer strategies for implantable bioelectronics. *IEEE Reviews in Biomedical Engineering*, 10:136–161, 2017.

[3]   A. Ahnood, K.E. Fox, N.V. Apollo, *et al.* Diamond encapsulated photovoltaics for transdermal power delivery. *Biosensors and Bioelectronics*, 77:589–597, 2016.

[4]   M. Ashraf and N. Masoumi. A thermal energy harvesting power supply with an internal startup circuit for pacemakers. *IEEE Transactions on Very Large Scale Integration (VLSI) Systems*, 24(1):26–37, 2015.

[5]   P. Assmann and Q. Summerfield. The perception of speech under adverse conditions. In *Speech processing in the auditory system*, New York, NY: Springer New York, 231–308, 2004.

[6]   S. Ayazian, V.A. Akhavan, E. Soenen, and A. Hassibi. A photovoltaic-driven and energy-autonomous CMOS implantable sensor. *IEEE Transactions on Biomedical Circuits and Systems*, 6(4):336–343, 2012.

[7] A.A. Babadi, S. Bagheri, and S.B. Abdul Hamid. Progress on implantable bio-fuel cell: Nano-carbon functionalization for enzyme immobilization enhancement. *Biosensors and Bioelectronics*, 79:850–860, 2016.

[8] A. Cadei, A. Dionisi, E. Sardini, and M. Serpelloni. Kinetic and thermal energy harvesters for implantable medical devices and biomedical autonomous sensors. *Measurement Science and Technology*, 25(1):012003, 2013.

[9] A.P. Chandrakasan, N. Verma, and D.C. Daly. Ultralow-power electronics for biomedical applications. *Annual Review of Biomedical Engineering*, 10, 2008.

[10] J. Charthad, M.J. Weber, T.C. Chang, and A. Arbabian. A mm-sized implantable medical device (IMD) with ultrasonic power transfer and a hybrid bi-directional data link. *IEEE Journal of Solid-State Circuits*, 50(8): 1741–1753, 2015.

[11] J.-F. Chen, C.-L. Chun, and Y.-Jr Hung. Mirror-assisted interdigitated back-contact CMOS photovoltaic devices for powering subcutaneous implantable devices. In *2015 International Symposium on Next-Generation Electronics (ISNE)*, pages 1–4. Piscataway, NJ: IEEE, 2015.

[12] Z. Chen, M.-K. Law, P.-I. Mak, and R.P. Martins. A single-chip solar energy harvesting IC using integrated photodiodes for biomedical implant applications. *IEEE Transactions on Biomedical Circuits and Systems*, 11(1):44–53, 2016.

[13] A. Cuadras, M. Gasulla, and V. Ferrari. Thermal energy harvesting through pyroelectricity. *Sensors and Actuators A: Physical*, 158(1):132–139, 2010.

[14] C. Dagdeviren, B.D. Yang, Y. Su, *et al.* Conformal piezoelectric energy harvesting and storage from motions of the heart, lung, and diaphragm. *Proceedings of the National Academy of Sciences*, 111(5):1927–1932, 2014.

[15] A.V. Dan Tchin-Iou and B.G. Min. Design of the solar cell system for recharging the external battery of the totally-implantable artificial heart. *The International Journal of Artificial Organs*, 22(12):823–826, 1999.

[16] R. Das, F. Moradi, and H. Heidari. Biointegrated and wirelessly powered implantable brain devices: A review. *IEEE Transactions on Biomedical Circuits and Systems*, 14(2):343–358, 2020.

[17] J.P. DiMarco. Implantable cardioverter–defibrillators. *New England Journal of Medicine*, 349(19):1836–1847, 2003.

[18] L. Halámková, J. Halámek, V. Bocharova, A. Szczupak, L. Alfonta, and E. Katz. Implanted biofuel cell operating in a living snail. *Journal of the American Chemical Society*, 134(11):5040–5043, 2012.

[19] K.O. Htet, R. Ghannam, Q.H. Abbasi, and H. Heidari. Power management using photovoltaic cells for implantable devices. *IEEE Access*, 6: 42156–42164, 2018.

[20] Y.-Jr Hung, M.-S. Cai, J.-F. Chen, *et al.* High-voltage backside-illuminated CMOS photovoltaic module for powering implantable temperature sensors. *IEEE Journal of Photovoltaics*, 8(1):342–347, 2017.

[21] G.-T. Hwang, M. Byun, C. K. Jeong, and K. J. Lee. Flexible piezoelectric thin-film energy harvesters and nanosensors for biomedical applications. *Advanced healthcare materials*, 4(5):646–658, 2015.

[22] J. Katic, S. Rodriguez, and A. Rusu. A high-efficiency energy harvesting interface for implanted biofuel cell and thermal harvesters. *IEEE Transactions on Power Electronics*, 33(5):4125–4134, 2017.

[23] A. Khan, Z. Abas, H.S. Kim, and I.-K. Oh. Piezoelectric thin films: An integrated review of transducers and energy harvesting. *Smart Materials and Structures*, 25(5):053002, 2016.

[24] L.W. Kleiner, J.C. Wright, and Y. Wang. Evolution of implantable and insertable drug delivery systems. *Journal of Controlled Release*, 181:1–10, 2014.

[25] T. Laube, C. Brockmann, R. Buß, *et al.* Optical energy transfer for intraocular microsystems studied in rabbits. *Graefe's Archive for Clinical and Experimental Ophthalmology*, 242(8):661–667, 2004.

[26] J.Y. Lee, H.Y. Shin, S.W. Kang, C. Park, and S.W. Kim. Improvement of electrical properties via glucose oxidase-immobilization by actively turning over glucose for an enzyme-based biofuel cell modified with DNA-wrapped single walled nanotubes. *Biosensors and Bioelectronics*, 26(5):2685–2688, 2011.

[27] X. Liang, R. Ghannam, and H. Heidari. Wrist-worn gesture sensing with wearable intelligence. *IEEE Sensors Journal*, 19(3):1082–1090, 2018.

[28] C. Liu, Y. Zhang, and X. Liu. Circularly polarized implantable antenna for 915 MHz ISM-band far-field wireless power transmission. *IEEE Antennas and Wireless Propagation Letters*, 17(3):373–376, 2018.

[29] L. Lu, Z. Yang, K. Meacham, *et al.* Biodegradable monocrystalline silicon photovoltaic microcells as power supplies for transient biomedical implants. *Advanced Energy Materials*, 8(16):1703035, 2018.

[30] K. MacVittie, J. Halámek, L. Halámková, *et al.* From "cyborg" lobsters to a pacemaker powered by implantable biofuel cells. *Energy & Environmental Science*, 6(1):81–86, 2013.

[31] V. Nabaei, R. Chandrawati, and H. Heidari. Magnetic biosensors: Modelling and simulation. *Biosensors and Bioelectronics*, 103:69–86, 2018.

[32] M.H. Ouda, M. Arsalan, L. Marnat, A. Shamim, and K.N. Salama. 5.2-GHz RF power harvester in 0.18-/spl mu/m CMOS for implantable intraocular pressure monitoring. *IEEE Transactions on Microwave Theory and Techniques*, 61(5):2177–2184, 2013.

[33] S. Ozeri, D. Shmilovitz, S. Singer, and C.-C. Wang. Ultrasonic transcutaneous energy transfer using a continuous wave 650 kHz Gaussian shaded transmitter. *Ultrasonics*, 50(7):666–674, 2010.

[34] S. Park, X. Guan, and Y. Kim, *et al.* PVDF-based piezoelectric microphone for sound detection inside the cochlea: Toward totally implantable cochlear implants. *Trends Hear*, 22:1, 2018.

[35] A. Rashidi, N. Yazdani, and A.M. Sodagar. Fully-integrated, high-efficiency, multi-output charge pump for high-density microstimulators. In *2018 IEEE Life Sciences Conference (LSC)*, pages 291–294. Piscataway, NJ: IEEE, 2018.

[36] F.R. Rezai. Dental treatment of patient with a cardiac pacemaker: Review of the literature. *Oral Surgery, Oral Medicine, Oral Pathology*, 44(5):662–665, 1977.

[37] R. Riahi, A. Tamayol, S.A.M. Shaegh, A.M. Ghaemmaghami, M.R. Dokmeci, and A. Khademhosseini. Microfluidics for advanced drug delivery systems. *Current Opinion in Chemical Engineering*, 7:101–112, 2015.

[38] M.I. Rudnicki, R.H. Chesworth, and L.T. Harmison. Design of a *sup*238 Pu-powered implantable thermal engine for heart assist devices. Technical report, Aerojet Nuclear Systems Co., Azusa, CA, 1970.

[39] M. Schormans, V. Valente, and A. Demosthenous. Practical inductive link design for biomedical wireless power transfer: A tutorial. *IEEE Transactions on Biomedical Circuits and Systems*, 12(5):1112–1130, 2018.

[40] J. Schwefel, R.E. Ritzmann, I.N. Lee, *et al.* Wireless communication by an autonomous self-powered cyborg insect. *Journal of the Electrochemical Society*, 161(13):H3113, 2014.

[41] M. Shen, W. Li, M.-Y. Li, *et al.* High room-temperature pyroelectric property in lead-free BNT-BZT ferroelectric ceramics for thermal energy harvesting. *Journal of the European Ceramic Society*, 39(5):1810–1818, 2019.

[42] B. Shi, Z. Li, and Y. Fan. Implantable energy-harvesting devices. *Advanced Materials*, 30(44):1801511, 2018.

[43] Q. Shi, T. Wang, and C. Lee. MEMS-based broadband piezoelectric ultrasonic energy harvester (PUEH) for enabling self-powered implantable biomedical devices. *Scientific Reports*, 6:24946, 2016.

[44] Y.-C. Shih, T. Shen, and B.P. Otis. A 2.3μw wireless intraocular pressure/temperature monitor. *IEEE Journal of Solid-State Circuits*, 46(11): 2592–2601, 2011.

[45] K. Shoji, Y. Akiyama, M. Suzuki, N. Nakamura, H. Ohno, and K. Morishima. Biofuel cell backpacked insect and its application to wireless sensing. *Biosensors and Bioelectronics*, 78:390–395, 2016.

[46] P. Song, S. Kuang, N. Panwar, *et al.* A self-powered implantable drug-delivery system using biokinetic energy. *Advanced Materials*, 29(11):1605668, 2017.

[47] K. Stingl, K.U. Bartz-Schmidt, D. Besch, *et al.* Artificial vision with wirelessly powered subretinal electronic implant alpha-IMS. *Proceedings of the Royal Society B: Biological Sciences*, 280(1757):20130077, 2013.

[48] T. Tokuda, T. Ishizu, W. Nattakarn, *et al.* 1 mm^3-sized optical neural stimulator based on CMOS integrated photovoltaic power receiver. *AIP Advances*, 8(4):045018, 2018.

[49] L.-G. Tran, H.-K. Cha, and W.-T. Park. RF power harvesting: A review on designing methodologies and applications. *Micro and Nano Systems Letters*, 5(1):14, 2017.

[50] J.D. Weiland, W. Liu, and M.S. Humayun. Retinal prosthesis. *Annual Review of Biomedical Engineering*, 7:361–401, 2005.

[51] C.T. Wentz, J.G. Bernstein, P. Monahan, A. Guerra, A. Rodriguez, and E.S. Boyden. A wirelessly powered and controlled device for optical neural control of freely-behaving animals. *Journal of Neural Engineering*, 8(4):046021, 2011.

[52] E.-J. Yoon, J.-T. Park, and C.-G. Yu. Thermal energy harvesting circuit with maximum power point tracking control for self-powered sensor node

applications. *Frontiers of Information Technology & Electronic Engineering*, 19(2):285–296, 2018.

[53]   Y. Yu, H. Sun, H. Orbay, *et al.* Biocompatibility and in vivo operation of implantable mesoporous PVDF-based nanogenerators. *Nano Energy*, 27: 275–281, 2016.

[54]   J. Zhao, R. Ghannam, M. Yuan, H. Tam, M. Imran, and H. Heidari. Design, test and optimization of inductive coupled coils for implantable biomedical devices. *Journal of Low Power Electronics*, 15(1):76–86, 2019.

[55]   M. Zhou, L. Deng, D. Wen, L. Shang, L. Jin, and S. Dong. Highly ordered mesoporous carbons-based glucose/O_2 biofuel cell. *Biosensors and Bioelectronics*, 24(9):2904–2908, 2009.

Chapter 3

Kinetic energy harvesting in implantable applications

By capturing the energy from body motions and turning it into useful electrical power, kinetic energy harvesting is one of the most promising methods for powering implantable medical devices. It provides a sustainable solution. As the need for durable, minimally intrusive medical equipment increases, kinetic energy harvesters offer a substitute for conventional batteries, lowering the frequency of repairs and improving patient comfort. Kinetic energy harvesting uses the heartbeat, breathing, and general motion of the body to provide a steady power source for a range of implanted devices, such as pacemakers, biosensors, and neurostimulators.

This chapter investigates the fundamentals, workings, and possibilities of kinetic energy harvesting in relation to implanted medical devices, looking at different approaches and how well they work for diverse uses. Kinetic energy harvesting's basic concepts are described in Section 3.1, which also describes how mechanical energy from bodily motions is transformed into electrical energy. The many forces at work – such as motion, compression, and vibration – are covered in this part along with the idea of energy conversion efficiency, which is essential to guaranteeing the usefulness of kinetic harvesters in low-energy bodily situations.

Section 3.2 addresses piezoelectric energy generators (PEGs) that make use of piezoelectric materials, which generate an electrical charge under mechanical stress. PEGs are especially suitable for recording repetitive motions of low frequency, such as heartbeats; the energy output is enough for a low-energy implant. This section focuses on material and design aspects of the piezoelectric devices, on how to optimize power generation while preserving the resilience of the body and its biocompatibility.

Triboelectric energy generators (TREGs), which create energy by the triboelectric effect – in which electrical charges are produced by the frictional contact between two materials – are discussed in Section 3.3. For implantation in regions of continuous, low-amplitude movement, TREGs are perfect because they are very good at absorbing energy from minute vibrations and even mild bodily movements. This section discusses the components, structural layouts, and efficiency elements that affect triboelectric generator performance in implanted environments.

Electromagnetic (EM) Generators, which are discussed in Section 3.4, generate electrical currents by means of the relative motion of coils and magnets. In settings where there are regular, bigger movements, like the heart or large muscle groups, EM generators work very well because they can extract a significant amount of energy from the tissues' inherent motion. In order to guarantee sufficient power generation within the limitations of implant size and placement, this section addresses the design issues for implantable EM generators, including the optimization of the coil and magnet configurations.

We move on to electrostatic (ES) generators in Section 3.5, which use the flow of electric charges between variable capacitance structures to produce energy. Miniaturized implants can benefit from ES generators because of their small size and potential incorporation into microelectromechanical systems (MEMS). The design and operating settings that optimize energy output while reducing wear and deterioration are the main topics of this section's discussion of the fundamentals of ES generating.

Section 3.6 presents adaptive kinetic energy harvesting (AKEH), a technique that optimizes energy output under various physiological conditions by combining many energy harvesting techniques or adjusting to various movement patterns. Because adaptive systems may react to shifts in movement patterns, including going from low to high activity levels, they are flexible options for implants in a range of physiological settings. The adaptive processes, hybrid designs, and sophisticated control systems that enable kinetic harvesters to optimize power output across a variety of body movements are covered in this section.

In summary, this chapter offers a thorough analysis of the implanted kinetic energy harvesting systems. Kinetic energy harvesting provides a way to develop self-sustaining medical gadgets that improve patient care and lessen dependency on conventional batteries by comprehending the many techniques and tailoring their design for particular bodily situations. In line with the objectives of minimally invasive, long-lasting, and patient-friendly healthcare solutions, these technologies will be essential in developing the next generation of energy-autonomous implants as research advances.

3.1   Principles of kinetic energy harvesting

The dynamical models of vibration energy harvesters (VEHs) are further examined in this part, with an emphasis on the intricate relationships between the mechanical and electrical domains in these systems. The way that VEHs transform mechanical vibrations into electrical power is the basis for their broad classification. There are two main types of VEHs: those that use the inertial force produced by the motion of a proof mass m (shown in Figure 3.1(b)) and those that apply an external force directly to the harvester (shown in Figure 3.1(a)). Inertial harvesters are very useful for microscale energy collecting applications since they only need one connection point to a vibrating structure, making integration easier [2,16].

The following notations are employed to model the kinetic harvesting system. The external driving force is denoted as $F(t)$. The inertial case is represented as $-m\ddot{y}$. The over-dot indicates the derivative with regard to time, whereas the variable $y(t)$ indicates vibrations. The relative displacement between the reference frame and the proof mass is determined by $z(t)$, and the potential energy is represented by $U(z)$, which for a linear system reduces to $\frac{1}{2}kz^2$, where k is the spring stiffness. The symbol

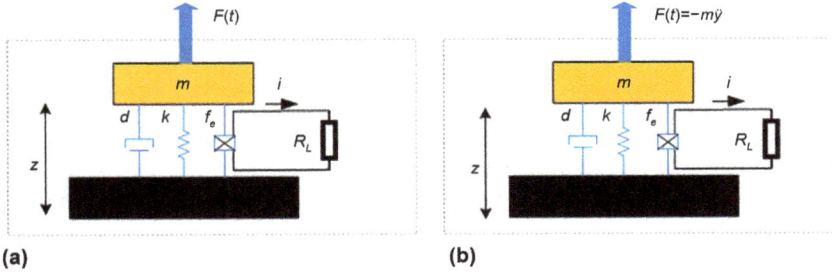

*Figure 3.1   Dynamical model of the kinetic energy harvesters: (a) with direct force
as well as (b) with inertial force*

for parasitic damping is d, and the symbol for the electrical restoring force originating
from the transduction mechanism is f_e. The load through which the generated current
i passes is represented by the resistance R_L [2,16].

The dynamical model developed by Williams and Yates says that the conversion
force, f_e, may be expressed as a force proportional to the relative velocity $f_e = -d_e \ddot{z}$,
which exerts a viscous damping effect. Note, however, that the electrical restoring
force f_e can, in general, be an elaborate function depending on acceleration, velocity, and mass displacement. One of the many interactions between the electrical and
mechanical subsystems, which this approximation neglects, despite its advantage, is
the feedback effect of the electrical load on the mechanical dynamics [16].

By including lumped parameters representing the electrical domain, the model
is extended as shown in Figure 3.1(b) in order to attain an even closer knowledge of
the dynamics of a VEH. Using this approach, the coupled governing (3.1) and (3.2)
of a one-degree-of-freedom vibration-driven generator may be derived by marrying
Kirchhoff's voltage law with Newton's second law [16]:

$$m\ddot{z} + d\dot{z} + \frac{dU(z)}{dz} + \alpha V = -m\ddot{y} \tag{3.1}$$

$$\dot{V} + (\omega_c + \omega_i)V = \omega_c \lambda \dot{z} \tag{3.2}$$

The motion dynamics of the inertial mass are described by (3.1), while the
dynamics of the connected electrical circuit are captured by (3.2). The electromechanical coupling factor is denoted by α, whereas V stands for the voltage produced
across the electrical resistance R_L. ω_c, which represents a high-pass filter effect unique
to the selected transduction approach, is the electrical circuit's characteristic cutoff
frequency. By using the formula $\omega_c = \frac{1}{\tau}$, this parameter is connected to the electrical circuit's characteristic time constant. Likewise, ω_i indicates the system's internal
resistance R_i, and λ represents the electromechanical conversion factor, which is
contingent upon the harvester's particular design and structure [16].

The potential energy function $U(z)$ can be represented as follows in the linear
case, which we concentrate on for simplicity's sake, shown in (3.3) [16]:

$$U(z) = \frac{1}{2}kz^2 \tag{3.3}$$

Due to the linear behavior of piezoelectric and EM harvesters, this model framework will be used mostly for their analysis. In contrast, because of the fluctuating force between ES plates, ES harvesters are intrinsically nonlinear. Consequently, a more intricate treatment is needed to completely represent the nonlinear dynamics of ES harvesters, even though the model presented here offers insightful information about piezoelectric and EM systems.

Equation (3.1) indicates that the vibration acceleration can be used to characterize the vibration force imparted to the harvester. More generally, this randomness in vibrational force can be explained by the fact that most, if not all, observable vibrations in real-world applications have unpredictable properties. The set of equations controlling the harvester's behavior thus becomes a set of stochastic differential equations. Named for the French physicist who proposed it in 1908 to explain Brownian motion, this collection of equations is commonly referred to as the Langevin equation set. Deterministically determining the solutions to these equations is impossible due to the stochastic nature of the force term. On the contrary, they can only be obtained in a statistical framework, provided that we have information on certain statistical aspects, such as moments and probability distributions, that characterize the random force. In general, specific methods from stochastic process theory are used in the study of stochastic differential equations. It is necessary to interpret any quantities of importance in this context, like the harvester's electric power production, as statistical averages of comparable quantities. By obtaining temporal averages of the solutions obtained from the dynamic equations, these numbers are ascertained. We use this statistical viewpoint in the parts that follow to provide light on how well the harvester performs when subjected to random vibrational forces. In conclusion, the stochastic nature of the vibrational force necessitates a probabilistic method of solving the differential equations, whereby significant solutions are obtained through time-averaged averaging. This method gives us a statistically significant description of the system's performance by capturing the expected behavior of the system under intrinsically random vibrational inputs [2,5].

3.2   Piezoelectric energy generator

PEG involves converting mechanical energy to electrical energy. The direct piezoelectric effect is well-suited for power harvesting that will induce a piezoelectric potential, attributing to the positive and negative charges of a polar surface if an external force is applied on the piezoelectric material [14]. This theory of the piezoelectric energy harvester is shown in Figure 3.2(a) and (b). The schematic and the structure of a piezoelectric energy harvester are shown in Figure 3.2(c), while the relative equivalent circuit is shown in Figure 3.2(d). The following coupled equation describes the piezoelectric effect:

$$S = S^E \cdot T + d^t \cdot E \tag{3.4}$$

$$D = d \cdot T + \epsilon^T \cdot E \tag{3.5}$$

where S is the strain tensor, T is the stress tensor, E is the electric field, D is the electric displacement, S^E is the compliance under a zero or constant electric filed, ε^T is the

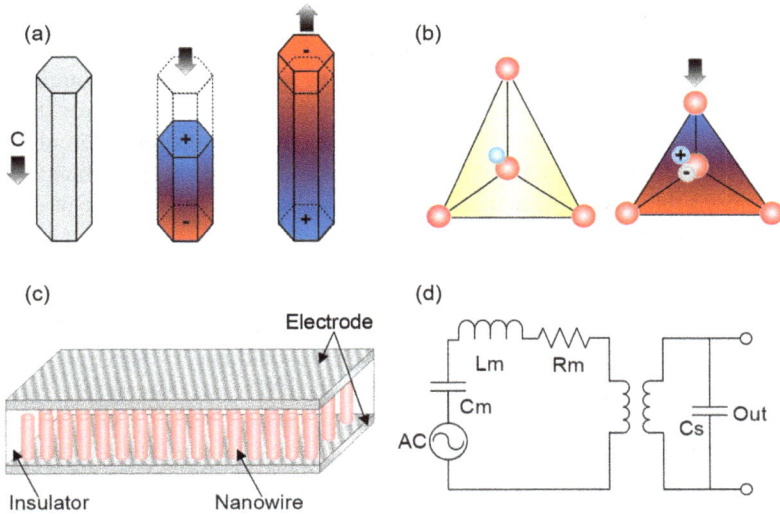

Figure 3.2   *Schematic diagram, mechanism, and structure of the triboelectric energy harvester. (a)–(d) The schematic diagram of triboelectric nanogenerator (TENG) structures and mechanisms. (e) A multilayer TENG structure with keel, sponge, and texture surfaces*

dielectric permittivity under a zero or constant stress, and d and dt are the direct and reverse piezoelectric coefficients [7].

The performance of piezoelectric systems is based on the characteristics of the piezoelectric material. The first implantable piezoelectric power system dates back to 1980 [8]. To date, the most common types of piezoelectric material are zirconate titanate (PZT), zinc oxide (ZnO), and polyvinylidene fluoride (PVDF) [14]. The first implantable PEG was fixed to a dog's ribs in 1984 [14]. Spontaneous breathing led to a PVDF cell producing 18 V and 17 μW of output peak voltage and power. PVDF material is advantageous since it is flexible and biodegradable, making it suitable for wearable and implantable applications [10].

PVDF material has been widely used in implantable devices. In Yanhao's work, the PVDF piezoelectric generator was embedded with a polydimethylsiloxane package, and the whole package was implanted into the rodent muscle. The stability of output was tested with an operating duration of 5 days. The average open circuit voltage (V_{oc}) and short circuit current (I_{oc}) were 3.8 V and 3.5 μA [17]. In 2015, a novel flexible and implantable PVDF PEG with capacitor storage was proposed with a size of 56 mm × 25 mm × 200 μm. It was tested both in vitro and in vivo. For the in vitro case, the maximum power output (P_{max}), V_{oc} and I_{sc} were 0.681 μW, 10.3 V and 400 nA. As for the in vivo study, the maximum current and voltage were 1.5 V and 300 nA when the PEG was attached to the heart of a male domestic porcine. The output power became 30 nW after 700 ms duration with a 70 bpm heart rate. This implantable PEG has shown the potential as a power source for low-power implantable electronic devices in the future [19].

Another PVDF piezoelectric study showed the in vitro and in vivo output power of 2.3 μW and 40 nW, respectively [3]. In 2016, other PEG materials were described in Khan *et al.* [10] and Shi *et al.* [15]. Similarly, the output power for ZnO and PZT are in the nW and μW scale [10,15]. ZnO material was used for encapsulation with textiles to convert the wasted mechanical energy into electric energy. Due to its quartzite crystal structure, PEG in ZnO can work as a high-frequency resonator in MEMS or nanoelectromechanical systems (NEMS). The PZT is more commonly used compared with the other materials because of its lighter weight. Although PZT suffers from a toxic nature, it still validates in vitro and in vivo studies [10].

A MEMS-based broadband piezoelectric ultrasonic energy harvester was previously developed to power implantable biomedical devices. The system was able to generate output power of 1.47 μW (without tissue) and 0.047 μW (with tissue), respectively [15]. Furthermore, this technique was FDA approved [1]. To avoid cytotoxicity of the constituent materials and immune response, the PZT device could be encapsulated by biocompatible materials. The power density of the PZT mechanical energy harvester on the bovine heart in vitro demonstration could reach 1.2 μ W/cm^2 and the output peak voltage is as large as 8.1 V [4]. Furthermore, piezoelectric sensors could harness environmental vibrations and convert them into electrical energy. For example, a 15-mm diameter PZT disk is involved to transfer ultrasonic transcutaneous energy and power the internal unit under a piece of pork muscle skin [1,12]. A power of 100 mW with 39.1% efficiency is successfully transferred at 650 kHz and 5 mm distance. The PZT material was utilized because of its high acoustic impedance. The acoustic matching layers were installed on the active surface of PZT material, and the backside was left open to allow ultrasonic escape, which massively improves the coupling energy into tissue. A Gaussian excitation of the transmitter approximation is applied to implement the ultrasonic to overcome the limitation of uniform excitation [1,12].

Therefore, piezoelectric energy harvesters can be designed to be flexible and small to meet human body vibration requirements. However, PZT is still the most popular piezoelectric material in recent devices, which is brittle and toxic. Thus, more research is needed into advanced materials that are more flexible. Another limitation of piezoelectric transducers is that they provide an AC voltage. Thus, an interface circuit is required to convert AC to DC electricity. This process increases system complexity, cost, and reduces the overall system efficiency.

3.3   Triboelectric energy generator

The TREG is based on the triboelecficition effect and ES induction, which occurs at the interface between the dielectric contacts [9]. When the two friction layers are moved to touch the contacts, triboelectric charges are delivered into the surface of the friction layers [9,20,21]. Once the friction layers are contacted, the same quantities of carriers will be induced. If the friction layers are separated, the ES field will be generated by the charge carriers, hence it is able to drive the electrons in the external load in the circuitry. Based on the structure specification, the operation mode of the TENG can be classified into four categories – Vertical contact-separation mode, Single Electrode mode, Lateral Sliding mode, and Freestanding mode – which is shown in Figure 3.3(a)–(d). The novel structure of the TENG is shown in Figure 3.3(e) with modified texture in the surface and keel and sponge structure. The conventional space-like shim and spring is widely applied in vertical contact-separation mode, and this is difficult in encapsulations in in vivo tests. The new technique with keel and sponge structure can replace the conventional spacers in TENG to enhance the energy output and stability [9]. The output performance can also be improved by the surface texture in use of polishing, inductive plasma etching and corona discharge technique [9].

A few TENGs were applied in the implantable system but it is still promising to be applied in the biomedical implantable system. In Zhang *et al.*, in 2014, the 400 μm separation is between the two contacts, and the whole TENG is encapsulated by the polydimethylsiloxane (PDMS) to provide flexibility and comfort. The TENG is implanted between e diaphragm and the liver in a rat and is tested in in vivo experiment. The energy is harvested in cycle according to the periodic breath of the rat. The output power is 8.44 mW/m^2, and the V_{oc} as well as I_{sc} are 3.74 V and 0.14 μA, respectively [14,22]. In Zhang *et al.*, in 2016, the TENG was firstly implanted and

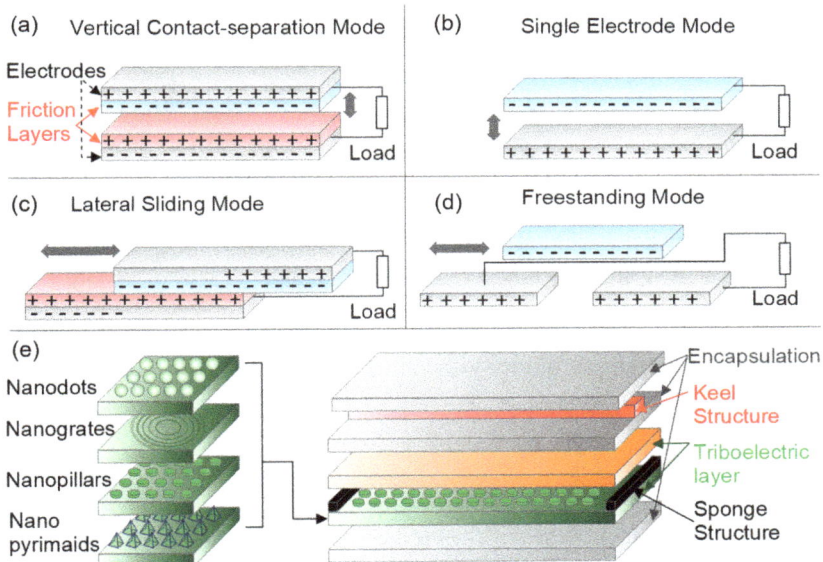

Figure 3.3   *Schematic diagram and mechanism of the biofuel cell (BC). (a) The schematic of the BC. (b) The several potential sources of malfunction for bioelectrodes after implantation. (c) The implantable BC in a living lobster with output voltage, output power density and implanting period [11]. (d) The implantable BC in a living snail with output voltage, output power density and implanting period [6]. (e) The implantable BC in a living rat with output voltage, output power density, and implanting period [18]*

tested in a living pig [23]. The TENG can supply the real-time cardiac monitoring system for 72 hours because of the biodegradable encapsulation of poly-lactic-co-glycolic acid (PLGA). Such a TENG can supply 14 V and 107 mW/m^2, respectively. In the device design and fabrication, a keel structure is integrated with a Ti strip on a Kapton substrate to improve the mechanical properties of the TENG [23].

3.4 Electromagnetic generators

Fundamentally, electromagnetic vibration energy harvesters (EMVEHs) work according to Faraday's law of induction. According to this law, which is fundamental to EM theory, an electromotive force, or voltage, will be induced within a conductive loop when a fluctuating magnetic flux passes through it. An electric current is essentially driven by an induced voltage created by any change in the magnetic flux connecting a conductor, provided that the circuit is closed. Generally, EMVEHs are constructed with at least a magnet and at least one conducting coil placed in such a way that they move with respect to each other. The relative motion is necessary because it keeps the atmosphere dynamic for energy generation through constantly changing the magnetic flux the coil goes through. It can be arranged such that the coil vibrates or oscillates near a fixed magnet, or a fixed coil is moved around a moving magnet. Due to the design of the device, it can convert the mechanical vibrations present in the environment into electrical energy, which may be useful. Applications in self-powered sensors and remote monitoring systems have sparked a lot of research interest in EMVEHs because of their capacity to capture ambient vibrations and convert them into electrical power. For low-energy devices, EMVEHs offer a sustainable and renewable power source by taking advantage of the relative motion between the coil and the magnetic components. The operational mechanisms of EMVEHs will be examined in the ensuing sections, along with the ways in which design elements like coil size, magnet strength, and relative displacement impact the devices' performance in practical settings.

As proposed by the concepts brought forward by Faraday's law of induction, for any alteration in the current of the system, there develops an induced voltage across the coil that can mathematically be represented as follows (3.6) [2]:

$$v_{cc} = -\frac{d\phi_{cc}}{dt} = -L\frac{di}{dt} \tag{3.6}$$

where L is the inductance of the coil, $\frac{d(i)}{dt}$ is the rate of change in current within time t, V_{cc} = induced voltage, and ϕ_{cc} = magnetic flux linkage. The negative sign shows that the induced voltage opposes the change in current, which agrees with the law of Lenz.

Also, due to the intrinsic resistance, R_L, of the coil, there is a voltage drop across it, according to Ohm's law. This resistance also draws part of the energy from the system as heat and contributes to the overall voltage. Combining these elements, the whole electrical governing equation for the EMVEH can be written as follows (3.6) [2]:

$$v(t) = \beta\frac{du(t)}{dt} - L\frac{di(t)}{dt} - r_L i(t) \tag{3.7}$$

where $v(t)$ denotes the total voltage across the harvester at time t, and $u(t)$ is the vibration or velocity of the moving component – which in most designs is a coil or magnet – inside the harvester, and β is a proportional constant related to some characteristics of the harvester. The term $L\frac{d(i)}{dt}$ characterizes the inductive hard voltage that reacts against the changes in current, whereas the term $R_L i(t)$ characterizes the voltage drop caused by the resistance of the coil.

This formula describes how the system's induced, impedance, and motion-generated voltages interact dynamically. For certain energy harvesting applications, the EMVEH's performance can be maximized by comprehending and modifying these elements. The basis for developing and evaluating more effective EMVEHs – which are becoming more and more important in supplying power to low-power electronic devices in a variety of settings – is laid by this thorough approach.

Two separate groups of EMVEHs are distinguished by Spree and Manoli's classification, which is shown in Figure 3.4.

The magnet in line configuration is one in which the coil's and the magnet's central axes line up. The magnet and coil's relative displacement directions are guaranteed to align with their respective central axes thanks to this configuration. The magnet's central axis is positioned perpendicular to the coil's central axis in this instance. An orthogonal relative displacement direction to the coil's central axis is produced by this configuration.

Along with putting out these divisions, Spree and Manoli also described the common architectures that fall within each group. Referenced and illustrated in Figure 3.4, their study illustrates the impact of these two configurations on the system. The magnetic flux intersecting the coil varies in each of the cases examined due to the relative movement between the coil and the magnet. Then a voltage is induced by this flux fluctuation.

Back iron is a term used in certain historic EMVEH designs. By focusing and channeling the magnetic field lines, the back iron increases the magnetic flux density that the coil experiences. By strategically optimizing the magnetic circuit, that is, by adding back iron, the electromechanical coupling factor, represented by the symbol β, can be raised. However, there is a major drawback to using a back iron. After reaching a saturation magnetic field, the material is unable to carry any more magnetic flux. It is crucial to carefully balance magnetic field intensity with the system's operational requirements because the anticipated performance improvements will not materialize if the system exceeds this saturation threshold.

3.5 Electrostatic generators

ESVEHs are basically based on the idea of changing the capacitance within a charged capacitor. When one electrode can move in relation to the other, the device has at least one capacitor by default, and this motion causes changes in capacitance. The relative movement between the electrodes caused by this capacitance fluctuation in turn makes energy generation easier. The structure of ES is shown in Figure 3.5 [2].

The concept of capacitance is introduced in the case of an isolated conductive structure that has an electric charge, and is shown in (3.8). Now, to measure the

Magnet in-line coil Architectures

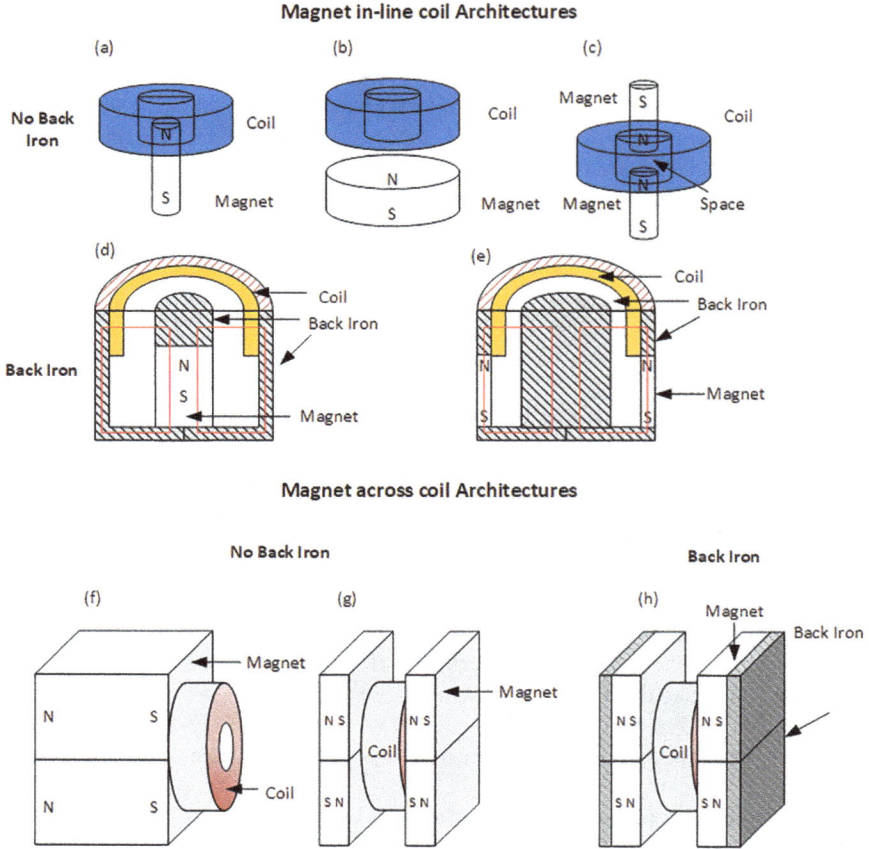

Figure 3.4 Electromagnetic harvester architectures are categorized according to the appearance of the back iron and the alignment of the coil and magnet. (a)–(e) are examples of "magnet in-line coil" structures, while (f)–(h) are examples of "magnet across coil" layouts. The inclusion or lack of back iron components is taken into consideration in further subdivision. A common axis is used by the coil and magnet in the "Magnet in-line coil" layout; other variants include extra structural elements like back iron or spacers. On the other hand, the "Magnet across coil" arrangement has the magnet and coil's central axis oriented perpendicularly, and some models also include a back iron to increase the concentration of magnetic flux

capacitance, the total charge is divided by the potential difference $V = |V_A - V_B|$ between the conducting structure [2].

$$C_0 = \frac{Q}{U} \tag{3.8}$$

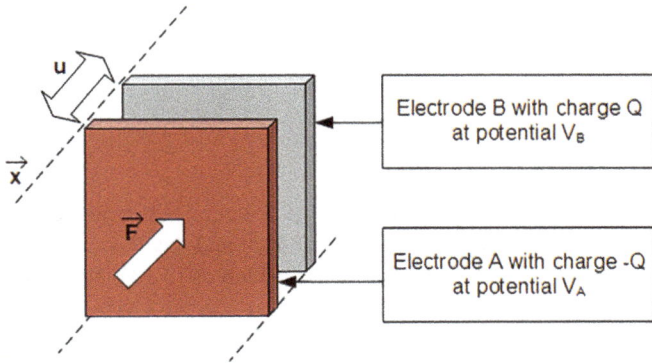

Figure 3.5   The simple structure of electrostatic capacitance fluctuation

where the magnitude of the difference in potential between two electrodes is used in calculating the difference in potential. The above statement emphasizes that dimensions, structure, shape, and relative location with respect to other conductors-particularly their electric mass-all affect capacitance. The higher the capacitance value, the better the structure for storing electrical charges. The capacitance of such structures usually ranges from a few picofarads pF to nanofarads nF [2].

The dependence of the system on the rate of change of charge may be expressed as follows $i = dQ/dt$. The simplest sort of capacitor to contemplate is a parallel-plane capacitor. It consists of two parallel electrodes. Such a capacitor will have a uniform electric field between the electrodes. This is so because the electrodes are much thinner than the lateral with of their plates and thus have uniform field distribution along the plates [2].

The general capacitance for this type of parallel-plane capacitor can be expressed as follows (3.9):

$$C_0(u) = \frac{\varepsilon_0 S_f}{u_0 + u} \tag{3.9}$$

where the variable distance ($u_0 + u$) between the electrodes in meters, (ε_0) vacuum permittivity, electrode surface area (S_f), and starting separation gap, respectively. From the above relationship, capacitance is inversely proportional to the distance between the electrodes; that is, as distance increases, the capacitance decreases and vice versa.

Equation (3.10) expresses an essential fundamental of electrostatics: the net outward flow of the electric field through a closed surface S is directly proportional to the total electric charge contained within that surface. In this mathematical illustration of Gauss's law, this fundamental law emphasizes the intrinsic linkage of electric flux and charge. Whatever the geometry of the surface, it proves to be a very solid base from which to study electric fields for different configurations, but, more important, it reveals that flux is dependent only on charge-enclosed.

$$\int_S \varepsilon_0 \vec{E} \cdot \vec{n} \, dS = \sum_S q' \tag{3.10}$$

where q' is the elementary charge in the volume that is bounded by S. The energy of a capacitor for a unit electric charge is known as the electric potential. E is the electric field, and dV is the variation of the elementary potential.

$$E = |\vec{E}|dV = Edu \tag{3.11}$$

From point A to point B in Figure 3.5, the potential difference can be evaluated as (3.12). Because the Electrical Field E is constant and the distance AB between the electrodes is equal to $u_0 + u$, (3.12) can be rewritten as (3.13). Normally, one of V_A or V_B is defined as a reference that is set to 0 [2].

$$V_B - V_A = -\int_A^B E\,dx \tag{3.12}$$

$$E = \frac{V_A - V_B}{u_0 + u} \tag{3.13}$$

It will be demonstrated that (3.8), (3.10), and (3.13), under the assumptions of a constant electric field and neglecting fringe effects, may be combined into (3.9). Neglecting fringe effects greatly simplifies the derivation and thus provides a more intuitive understanding of the dependence of capacitance on the physical structure of the capacitor. The energy from the ES generator can be regarded as $C_0(u)$ stored charge Q, which is shown in (3.14):

$$W = \frac{1}{2}QV = \frac{1}{2}C_0(u)V^2 = \frac{1}{2}\frac{Q^2}{C_0(u)} \tag{3.14}$$

In terms of operation mechanism, variable capacitors can be separated into mainly two varieties: the in-plane gap-closing configuration illustrated by Figure 3.6(b)–(e) and out-of-plane gap-closing designs exemplified by Figure 3.6(a), as well as Figure 3.6(c). For most design variations in capacitance, the out-of-plane gap-closing design allows, in counterpart, however, to preclude an overdisplacement, it is necessary to include a mechanical stopper. In contrast, the in-plane gap-closing method is a simpler and perhaps more reliable option because it operates without the need for mechanical stoppers, although it provides a somewhat lower range of capacitance variation. However, there are stability issues with the in-plane configurations, especially where large displacements occur [2,13].

Although the term "out-of-plane" would be more appropriate in this context to describe the nature of the relative motion of the electrode surfaces, the so-called "out-of-plane finger type" appearing in Figure 3.6(c) is frequently categorized as an in-plane gap closing structure due to its peculiar displacement [2,13]. Among them, the boss-type structure is especially fit for variable-amplitude displacement applications. Due to its structure, it is possible to calculate a maximum capacitance fluctuation for a specific bump pattern pitch, and this is helpful in applications requiring dynamic and accurate capacitance management [2,13].

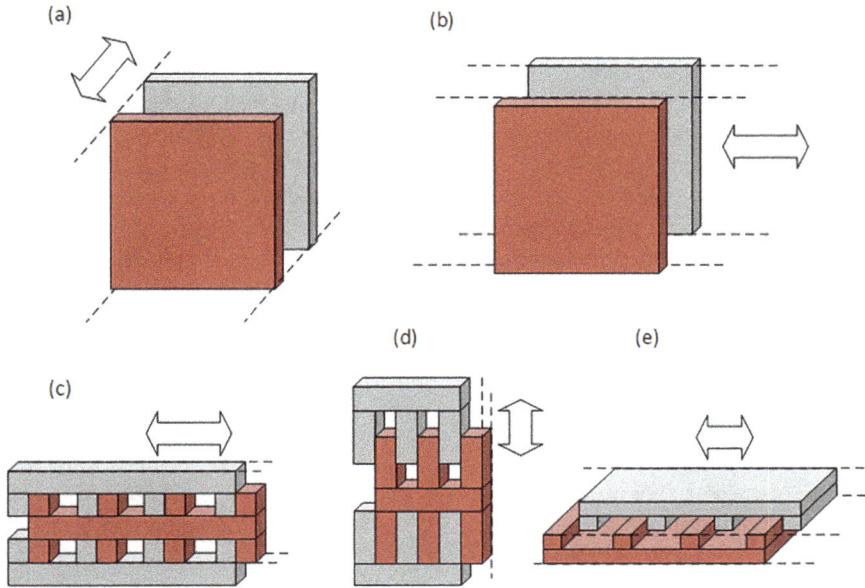

Figure 3.6   Variable configurations, along with their respective derived structures, can be grouped as (a) out-of-plane gap closing configuration, OP; (b) in-plane variable overlap surface arrangement, IP; (c) out-of-plane finger-type structure, OP; (d) in-plane finger-type design, IP; and (e) in-plane boss-type configuration, IP. Each one of these arrangements use different mechanical and geometrical principles to realize variable capacitance, and all these principles are tailored to meet various demands for a range of applications. These designs demonstrate variable capacitors' capability for adaptability and versatility in very specific functional requirements of modern electronic systems

3.6   Adaptive kinetic energy harvesting

Kinetic energy harvesting, long considered a potential alternative to finite-capacity batteries may power implanted devices. Conventional kinetic energy harvesting devices represent a static and rigid architecture, hence they are not suitable for implantable contexts since they cannot adapt to the extremely dynamic and changeable nature of human body movements. AKEH is an innovative breakthrough in this area that introduces technologies capable of dynamically adapting to the physiological parameters of the body. This chapter discusses some of the basic differences between adaptive and traditional kinetic energy harvesting to highlight the great improvements in the powering of implanted devices made possible by AKEH.

- Conventional kinetic energy harvesters are fixed-frequency designs that are optimized for specific mechanical motions, such as walking or pulse. When these systems are applied to implantable devices, several limitations have been evident despite success in controlled environments.

- Static frequency response: Classic harvesters are designed to resonate at a single frequency; this limits their ability to capture energy from the wide range of frequencies that many types of body motion variations can exhibit.
- Restricted adaptability: These systems cannot adapt to changes in posture, activity, or physiological status, which produce altered motion characteristics, for example, rest compared to exercise.
- Consequently, suboptimal energy gain takes place under suboptimal conditions; at resonance, body movements can really lower the energy conversion efficiency, leading to a variable supply of power.
- Miniaturization complexity: In general, conventional systems use supplementary mechanical parts like springs or stoppers to limit their movement. In addition, this complicates their miniaturization into small implantable devices.

AKEH combines dynamic tuning and real-time adjustment methods to address the disadvantages of conventional systems. This invention will have higher efficiency and reliability in the complex and changing environment of the human body.

- The tunable resonator or smart material in adaptive systems changes frequency according to the motion pattern. For example, during exercise, AKEH can tune its frequency to capture energy from higher-intensity movements. During rest, it can be tuned to small low-frequency movements such as breathing or cardiac movements.
- Higher output under various circumstances: Unlike conventional piezoelectric systems, AKEH generates consistent energy regardless of posture, activity, and/or motion change. This makes it more adaptive, losing less energy while providing a stable power output even in conditions that are not anticipated.
- Reduced reliance on mechanical stoppers: In traditional systems, the dependency relies on either a stopper or the geometry of the set to prevent overdisplacement. AKEH reduces this dependency by making self-regulating designs using smart material or programmable circuits, reducing mechanical wear and increasing the life span.
- Biomechanical integration: An adaptive system can be tuned to the mechanical characteristics of the implant site, such as tissue elasticity or the amplitude of motion. In this way, it is ensured that the capture is maximized while the normal functioning of the body is not interfered with.
- AKEH systems can implement different energy harvesting mechanisms, such as piezoelectric, EM, and ES, to realize a hybrid configuration able to adapt itself to the kind of motion detected.

The comparison Table 3.1 shows the difference between the conventional kinetic energy harvester and AKEH:

While AKEH is a big step forward, there are still issues in optimizing its speed and scalability. Further advances in microfabrication processes are needed to create adaptive systems that are small enough for widespread implanted use. Biocompatible and durable materials are in great demand for long-term successful implantation.

Table 3.1   *A comparison of a conventional kinetic energy harvester and an adaptive energy harvester*

Parameter	Conventional kinetic energy harvester	Adaptive kinetic energy harvester
Frequency	Fixed	Adaptive
Energy Efficiency	Efficient under specific condition	Consistently efficient across specific condition
Suitability for Motions	Limited	Highly adaptable
Design Complexity	Mechanically intensive	Smarter and self-regulating design
Body Motion Integration	Limited	Adaptive to biomechanics
Reliability	Subject to wear and drift	Reliable

Increasing the amount of energy collected per unit volume remains a top priority, especially in the powering of implants that demand high energy. Future research will probably be directed at the incorporation of artificial intelligence and machine learning into AKEH systems for predictive modifications based on user activity patterns. The combination of AKEH with other energy harvesting methods, such as thermal or biochemical harvesting, may also improve the energy autonomy of implantable devices.

References

[1]   K. Agarwal, R. Jegadeesan, Y.-X. Guo, and N.V. Thakor. Wireless power transfer strategies for implantable bioelectronics. *IEEE Reviews in Biomedical Engineering*, 10:136–161, 2017.

[2]   D. Briand, E. Yeatman, and S. Roundy, editors. *Micro Energy Harvesting*. New York: Wiley-VCH, 2015.

[3]   X. Cheng, X. Xue, Y. Ma, *et al.* Implantable and self-powered blood pressure monitoring based on a piezoelectric thinfilm: Simulated, in vitro and in vivo studies. *Nano Energy*, 22:453–460, 2016.

[4]   C. Dagdeviren, B.D. Yang, Y. Su, *et al.* Conformal piezoelectric energy harvesting and storage from motions of the heart, lung, and diaphragm. *Proceedings of the National Academy of Sciences*, 111(5):1927–1932, 2014.

[5]   C.W. Gardiner. Handbook of Stochastic Methods, 4th edn., Springer, Berlin, 2009.

[6]   L. Halámková, J. Halámek, V. Bocharova, A. Szczupak, L. Alfonta, and E. Katz. Implanted biofuel cell operating in a living snail. *Journal of the American Chemical Society*, 134(11):5040–5043, 2012.

[7]   T. Hehn and Y. Manoli. CMOS circuits for piezoelectric energy harvesters. *Springer Series in Advanced Microelectronics*, 38:21–40, 2015.

[8]   W. Heimisch, S. Hagl, K. Gebhardt, N. Mendler, and H. Meisner. Cyclic changes in ventricular wall geometry measured by implantable piezoelectric tubes. In *Physics in Medicine and Biology*, volume 25, pages 981–982. Publishing Ltd Dirac House, Bristol, 1980.

[9]   D. Jiang, B. Shi, H. Ouyang, Y. Fan, Z.L. Wang, and Z. Li. Emerging implantable energy harvesters and self-powered implantable medical electronics. *ACS Nano*, 14(6):6436–6448, 2020.

[10]   A. Khan, Z. Abas, H.S. Kim, and I.-K. Oh. Piezoelectric thin films: An integrated review of transducers and energy harvesting. *Smart Materials and Structures*, 25(5):053002, 2016.

[11]   K. MacVittie, J. Halámek, L. Halámková, *et al.* From "cyborg" lobsters to a pacemaker powered by implantable biofuel cells. *Energy & Environmental Science*, 6(1):81–86, 2013.

[12]   S. Ozeri, D. Shmilovitz, S. Singer, and C.-C. Wang. Ultrasonic transcutaneous energy transfer using a continuous wave 650 kHz Gaussian shaded transmitter. *Ultrasonics*, 50(7):666–674, 2010.

[13]   S. Roundy, P.K. Wright, and J. Rabaey. A study of low level vibrations as a power source for wireless sensor nodes. *Computer Communications*, 26(11):1131–1144, 2003.

[14]   B. Shi, Z. Li, and Y. Fan. Implantable energy-harvesting devices. *Advanced Materials*, 30(44):1801511, 2018.

[15]   Q. Shi, T. Wang, and C. Lee. MEMS based broadband piezoelectric ultrasonic energy harvester (PUEH) for enabling self-powered implantable biomedical devices. *Scientific Reports*, 6:24946, 2016.

[16]   C.B. Williams and R.B. Yates. Analysis of a micro-electric generator for microsystems. *Sensors and Actuators A: Physical*, 52(1–3):8–11, 1996.

[17]   Y. Yu, H. Sun, H. Orbay, *et al.* Biocompatibility and in vivo operation of implantable mesoporous PVDF-based nanogenerators. *Nano Energy*, 27: 275–281, 2016.

[18]   A. Zebda, S. Cosnier, J.-P. Alcaraz, *et al.* Single glucose biofuel cells implanted in rats power electronic devices. *Scientific Reports*, 3(1):1–5, 2013.

[19]   H. Zhang, X.-S. Zhang, X. Cheng, *et al.* A flexible and implantable piezoelectric generator harvesting energy from the pulsation of ascending aorta: in vitro and in vivo studies. *Nano Energy*, 12:296–304, 2015.

[20]   Q. Zhang, Q. Liang, Q. Liao, *et al.* Service behavior of multifunctional triboelectric nanogenerators. *Advanced Materials*, 29(17):1606703, 2017.

[21]   Q. Zhang, Z. Zhang, Q. Liang, *et al.* Green hybrid power system based on triboelectric nanogenerator for wearable/portable electronics. *Nano Energy*, 55:151–163, 2019.

[22]   Q. Zheng, B. Shi, F. Fan, *et al.* In vivo powering of pacemaker by breathing-driven implanted triboelectric nanogenerator. *Advanced Materials*, 26(33):5851–5856, 2014.

[23]   Q. Zheng, H. Zhang, B. Shi, *et al.* In vivo self-powered wireless cardiac monitoring via implantable triboelectric nanogenerator. *ACS Nano*, 10(7): 6510–6518, 2016.

Chapter 4
Thermal electric energy harvesting in implantable applications

Thermoelectric devices convert thermal energy into electricity by the Seebeck effect. The Seebeck effect is a thermoelectric phenomenon that involves converting a temperature difference into a voltage difference. This phenomenon occurs mainly in metals and semiconductors. Heating one end of a semiconductor causes a temperature difference, which enables carriers to diffuse from the hot to the cold ends of this semiconductor. The schematic of the thermal electric generator (TEG) is shown in Figure 4.1(a) and (b), which is the structure of the bimaterial cantilever [5].

4.1 Principles of thermal energy harvesting

Considering an n-type semiconductor as an example, due to the high concentration of electrons, the majority carriers will diffuse from the heated side to the cooler side in the semiconductor. The minority carriers (holes) will, in turn, move in the opposite direction. In open-circuit conditions, negative charges at the hot end and positive charges at the cold end are formed at either side of the semiconductor, which results in an electric field to appear inside the semiconductor. When the semiconductor reaches a stable state, the electromotive force caused by this temperature difference appears at both ends of the semiconductor, where the procedure is shown in Figure 4.1(c)–(f). Both p- and n-type semiconducting materials are required to cause current flow in a TEG. Thermal energy harvesting has been used in many sensing applications [2]. The thermoelectric effect can be described by the following equations [1]:

$$V_G = N \propto \Delta T \tag{4.1}$$

$$P_L = N^2 \alpha^2 \Delta T^2 \frac{R_L}{R_L + R_{in}} \tag{4.2}$$

where V_G is the output voltage, α is the Seebeck coefficient of the thermal material, N is the number of thermocouples, R_L is the loading resistance, R_{in} is the internal resistance of TEG, and PL is the output power. It is well known that the human body is an unlimited heat source, which leads to great potential in implantable energy harvesting.

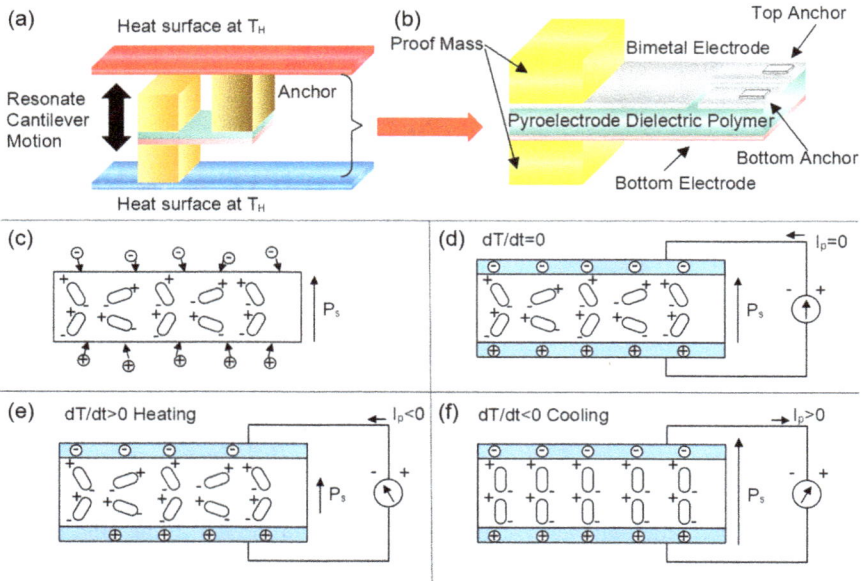

Figure 4.1 Schematic diagram and mechanism of the thermal electric generator
(TEG). (a) The schematic of the pyroelectric device in a bimaterial
cantilever structure. (b) Details of the construction of the capacitive
cantilever thermal energy converter and split anchor structures. (c)–(f)
The pyroelectric material used as dielectric in a capacitor [5]. As the
temperature of the pyroelectric generator is increased, the current in
the external circuit is compensates for the change in bound charge at
the edges of the crystal

4.2 Materials and methods

Conventional thermoelectric materials have been widely explored and employed in
real-world applications. Some of the most popular materials include the following:
bismuth telluride – best suited to low-temperature applications (300–500 K), with a
high Seebeck coefficient and a low thermal conductivity, widely used in commercial
thermoelectric systems, particularly for cooling; lead telluride (PbTe) – effective at
temperatures ranging from 500 to 900 K (it has a high efficiency because of its favor-
able Seebeck coefficient and modest thermal conductivity); silicon-germanium alloys
(SiGe) – used in high-temperature (over 1000 K) applications, such as space exploration, provides high thermal stability, but has poor thermoelectric figure of merit
(ZT) values when compared to other materials. According to the range of the temperature and toxicity, the bismuth telluride is advantageous in implantable applications.
The pyroelectric effect is inextricably related to materials possessing piezoelectric
capabilities. Indeed, all known types of pyroelectric crystals are induced to produce
an electric field by mechanical stress – a signature of the piezoelectric effect [8].

Of the 32 crystal classes recognized in crystallography, 20 can produce an electric field upon mechanical stress, which is characteristic of piezoelectricity. Of the 20 piezoelectric crystal classes, 10 are pyroelectric, meaning they are able to develop an electric field in response to time-varying temperature.

4.3   Thermoelectricity harvesting

It refers to the generation of electric outputs from temperature differences through varied methods. Thermoelectric generation is based on what is called the Seebeck Effect – a physical phenomenon named in recognition of its discoverer, German physicist Thomas Johann Seebeck. Basically, the effect consists of a phenomenon wherein an electromotive force is developed through conduction and by submitting said conductor to a temperature gradient in its junctions [8].

Thermoelectric devices are traditionally made from two types of semiconductors, n-type and p-type materials, often based on bismuth telluride or bismuth selenide. These semiconductors can be distinguished by the Seebeck coefficient, a measure of voltage produced per unit of temperature gradient. Here, the excess of negative charge carriers (electrons) means the n-type material possesses a negative Seebeck coefficient, while in the p-type material, the excess of positive charge carriers is known as holes will make it possess a positive Seebeck coefficient [8].

TEGs will work based on electrical series connection and thermal parallel connection. Junctions between two materials are alternately exposed to hot and cold thermal reservoirs often through electrical insulating layers. The heat that flows from the hot to the cold side diffuses charge carriers across semiconductors, electrons in n-type material flow to the cold side, while the holes in p-type material flow in the opposite direction. These carrier motions create an electric current that traverses an external circuit. The voltage developed is directly proportional to the temperature difference between the two materials and also to the difference between their Seebeck coefficients [8].

In fact, TEGs were quite inefficient up to the early 2000s due to the highly challenging material requirements for decent thermoelectric performance. The optimum material for thermoelectric devices should have high electrical and low thermal conductivity to easily carry on charge carriers and not dissipate the temperature across the device. Unfortunately, good electrical conductors, typically, like metals, turn out to be good thermal conductors, whereas good thermal insulators are poor electrical conductors, as in the case of ceramics. This intrinsic trade-off in material properties has generally limited the performance of TEGs.

Although significant improvement has occurred, there are still large areas of improvement for thermoelectric materials and device designs. In the case of nanostructuring, for example, promising avenues have been investigated toward reduced thermal conductivity without the penalty of increased ZT. Explorations of hybrid materials – that is, those that combine the metal-like electrical properties with the ceramic-like thermal properties – have led to potential breakthroughs in performance. Different other techniques employing Seebeck devices in the realm of energy harvesting in low-temperature regions have been widely varying, as represented in a bundle of papers such as [6]. A novel small greenhouse design for the expansion of the temperature gradient has been implemented in a series of trials. This setup increased the ambient temperature of the hot side of the thermoelectric device while the cold side of the device was in direct thermal contact with a heat sink buried in water, creating a proper and pronounced temperature difference. In optimizing the power generated, several designs in catch areas, electrical wiring systems, and construction designs were closely evaluated in detail. Under average ambient conditions of about 300 K, the greenhouse design was able to raise the hot side to 350 K and produced more than 14 mW of power [17].

A similar approach was taken toward soil-based applications in [15], where thermoelectric devices were embedded in a soil matrix to leverage natural temperature gradients. The earth worked as a thermal reservoir, the hot side facing sunlight, while the cold side was cooled by subsurface conduction or passive heat dissipation. This concept illustrated the ability of thermoelectric systems to adapt to low-temperature differences in a variety of environmental conditions. The first implantable thermoelectric power harvester was demonstrated in 1970 [10]. The most common semiconductor material for thermoelectric power harvesting is polycrystalline silicon germanium (poly-SiGe) and bismuth telluride (Bi–Te). However, due to its high ZT properties and room temperature fabrication capabilities, Bi–Te is mainly used for commercial applications [12]. Compared to poly-SiGe, Bi–Te can generate between 19 and 30 μW/mm^2 with a temperature difference of 27 and 70 K [20]. Thus, TEGs have many advantages that include their lightweight and flexibility. However, these energy harvesters are not ideal in environments that have a temperature similar to that of the human body [13].

4.4   Pyroelectricity energy harvesting

Pyroelectricity is the property of generating an electric potential in a crystalline material as a result of temporal variation in temperature. As claimed by [3,8,18], such materials inherently and stably possess dipole moments at every instant of the crystalline structure. Due to this continuous process of variation in temperature, dipoles within the crystal take up a particular direction, thus creating an overall electric potential. If the crystal is kept at a constant temperature, the internal depolarization field cancels the free charges on its surface, and the system is stabilized.

The principal weakness of all real systems intended to generate electricity from pyroelectric materials is the intrinsic reliance of these systems on time-varying temperatures to produce an electrical field. The former then calls for dynamic temperature cycling, which may overcomplicate both design and operation of a system when there is only a small or hardly controllable variation of temperature. However, thermoelectric devices are more adaptable because they generate energy using only a temperature gradient. This state is sometimes achieved passively by incorporating cooling mechanisms such as heat sinks or thermal dissipation structures, making them more flexible to a wide range of applications [8].

Although thermoelectric devices can maintain a constant temperature difference rather efficiently, they have inherently worse energy conversion efficiency compared to an ideal pyroelectric material. Pyroelectric devices are able to utilize fleeting thermal events that would normally remain unrecovered, including discontinuous process-related waste heat, when a variety of temperature control systems are used by them. It would reduce these operating issues with pyroelectric energy harvesters in your approach by integrating the latest features of new materials with high conductivity in the heat transfer mode or microscale temperature actuators [8]. The application settings are yet another significant factor that greatly influences the choice between pyroelectric and thermoelectric systems. Pyroelectric devices are more adapted for situations with natural or planned temperature changes, while under conditions of constant temperature gradients, thermoelectric systems come into their own. In the development of hybrid systems that incorporate capabilities from both of these technologies, further improvement in energy recovery may be possible for a wide range of dynamic situations.

In most practical situations, the temperature fluctuations that occur externally are too negligible to result in a decent voltage being induced across the pyroelectric devices. A method that overcomes this limitation in terms of the conversion of temperature gradients into time-varying temperature profiles utilize a working fluid that is exchanged between hot and cold reservoirs described by Sebald *et al.* Indeed, this approach involves coupling the system with an adequately designed thermodynamic cycles for maximum efficiency in energy harvesting. These often include precise management of thermal connection to external temperature reservoirs and dynamic modification of the electrical load, with the view of extracting energy at optimal instances in the cycle. The thermodynamic cycle used greatly influences the performance of the system since energy output and operating efficiency are changed, as Sebald *et al.* [11] show analytically.

One of the most promising thermodynamic cycles for pyroelectric energy harvesting is the synchronized switch damping on the inductor. The synchronized switch damping on inductor (SSDI) cycle utilizes the principles of power electronics, positioning an inductor at the device output to dynamically control the load. When the gadget reaches its minimum temperature, it inverts the load polarity, enhancing the energy-transfer process. According to Sebald *et al.* [11], with the right engineering tactics and material selection, this cycle can obtain efficiencies that are close to 50% of the Carnot limit, which is very high for thermal energy conversion systems.

The reason why the SSDI cycle is special is that it somehow manages to synchronize thermal and electrical dynamics in such a way as to achieve effective energy harvesting in conditions that are far from ideal. However, in practice, the implementation of such cycles is quite challenging. Examples of inductors that can be applied in implantable or small-sized devices and work continuously with high performance are currently under active research. The switching synchronization mechanisms at the precise thermal states should be equally accurate to eliminate the energy losses.

More innovation in materials science in the development of pyroelectric materials with better thermal and electrical coupling could significantly increase the feasibility and efficiency of SSDI-based systems. In addition, the use of advanced thermal management technologies, such as microfluidic heat exchangers, might enable the optimization of temperature gradients and the improvement of performance. The combination of SSDI cycles with hybrid energy harvesting technologies, such as merging pyroelectric systems with thermoelectric or piezoelectric harvesters, may also provide increased efficiency and variety in energy harvesting applications.

Ravindran *et al.* [9] discussed a more practical version of pyroelectric generators, thus overcoming fundamental issues in thermal energy harvesting. The architecture of the heat engine promotes thermal energy transfer between a hot and a cold reservoir by employing an oscillating diaphragm to produce continuous temperature fluctuations in the pyroelectric material. Although the exact rate of heating is not clearly mentioned in their analysis, the cooling part of the cycle demonstrates substantial thermal dynamics. For example, a maximum temperature difference of 79.5 K can be achieved in less than one second, which is a great improvement compared to the previous work, such as [19].

This novel strategy leads to significant performance improvements. The generator produces maximum open-circuit voltages of about 13 V and an average power density of 3.03 $\mu W/cm^2$, which is significantly higher than the output reported in [19]. These findings highlight the importance of improved thermal management tactics in maximizing the energy harvesting capabilities of pyroelectric materials.

Both studies [9,19] point to the stability of pyroelectric materials during thermal cycling, indicating that they function up to their Curie temperature. This crucial characteristic is the maximum working temperature at which the material loses its pyroelectric properties and is highly dependent on the composition of the material. For example, PVDF has a Curie temperature of approximately 165°C, while lead zirconate titanate (PZT) can withstand temperatures up to 350 °C, as stated by [19]. These high-temperature tolerances increase the application of pyroelectric generators, particularly in high-temperature situations such as industrial waste heat recovery or geothermal systems [11].

The importance of the oscillating diaphragm is underlined in the approach proposed by Ravindran *et al.* to ensure fast and periodic temperature modulation. Such an operational mode creates a dynamic thermal environment, enabling effective energy conversion due to efficient exploitation of the material response against regular thermal cycling. However, additional research is needed regarding the endurance of such mechanical components when considering long-time operation at high thermal loads.

Future improvements in pyroelectric energy harvesting may be related to the use of new materials with improved thermal and electrical properties, such as doped PZT or nanostructured composites. Moreover, the integration of oscillatory systems with state-of-the-art heat transfer methods, like microfluidic channels or phase-change materials, might lead to a significant enhancement of the heat exchange process, hence providing even higher power density. Amelioration of these factors might allow pyroelectric generators to become highly efficient, scalable systems for sustainable energy harvests in a wide range of applications [11].

4.5   Thermomagneticity energy harvesting

Thermomagnetic generators (TMGs) are based on the principle that the magnetization of certain materials changes with temperature. These temperature changes in magnetization produces changes in the magnetic field of such material, which can apply force to other nearby magnetic objects to do mechanical work or induce electrical currents in nearby conductive materials and hence deliver electrical power. This dual feature illustrates the versatility of TMGs in energy conversion applications.

The theoretical possibility of TMGs has been studied since the late 1940s, and several research articles pointed to the fact that their thermal-to-electrical energy conversion efficiency may be comparable to or even higher than that of state-of-the-art thermoelectric devices [4,8]. However, early experimental efforts were limited by the lack of strongly magnetic materials and materials with suitable thermomagnetic properties at about ambient temperatures. Advances in materials research have since overcome these challenges, and materials with better magnetic properties and Curie values compatible with practical operation conditions have been discovered.

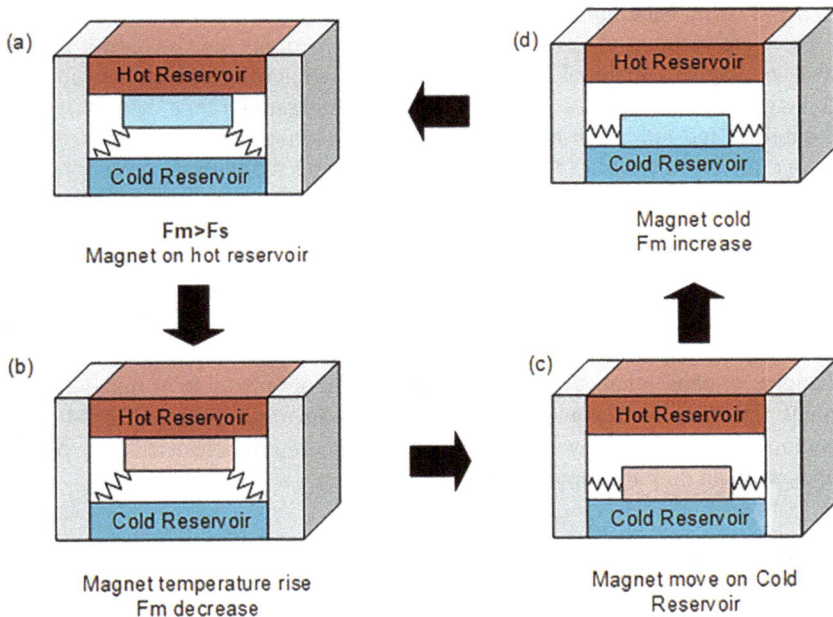

Figure 4.2   *A schematic of the most basic thermomagnetic cycle: (a) the magnet is on the hot reservoir, where the magnetic force Fm is greater than the restoring force Fs; (b) the magnet temperature rises and Fm decreases; (c) the magnet moves onto the cold reservoir; (d) the magnet becomes cold and Fm increases. This cycle illustrates the interaction between the magnetic force Fm and the restoring force Fs, showing how the thermomagnetic effect works. The process repeats itself due to the restoring force opposing Fm at several important positions, thus explaining the basic principles that drive thermomagnetic production and allowing for a better understanding of its operational mechanics [8].*

However, the subject has not received the same attention as other energy conversion technologies, and actual demonstrations supporting interesting theoretical predictions are few.

In fact, TMGs operate by repeatedly heating and cooling a magnetic material in contact with thermal reservoirs at different temperatures, which is shown in Figure 4.2. This process can be performed actively. The passive design is an evident implementation of a thermomagnetic generation cycle. In this device, the magnetization of a magnetic material varies cyclically as it switches periodically between contact with a hot and a cold reservoir. When the material is in contact with the cold reservoir, it cools by thermal conduction, which increases its magnetism. The magnetic force acting on the material is large enough to overcome a restoring force, such as one provided by a spring, and pull the material toward the hot reservoir. As long as the material is in contact with the hot reservoir, its magnetization decreases because of the heat. This reduced magnetic force allows the restoring force to dominate, which pulls the material back into the cold reservoir; the process repeats.

There are substantial obstacles to the real application. One key barrier is how to synthesize magnetic material with an ideal combination of characteristics, including large magnetization variation over an extended temperature range, appropriate thermal stability, and minimum hysteresis losses. In addition to this, the effective heat exchange process between the magnetic material and the thermal reservoirs is one of the open issues needed to reduce thermal losses and to increase efficiency energy conversion.

Further work in the development of such devices could be implemented with the use of advanced material systems, such as magnetic alloys based on rare-earth elements or nanostructured magnetic composites. In addition, new thermal management solutions might be developed, implementing, for example, microfluidic heat exchangers or phase-change materials to achieve an optimization of the temperature gradient. Hybrid systems could represent a further step toward this end: a TMG might be combined with other conversion principles, such as thermoelectric or piezoelectric devices, with an output adapted to the actual request.

4.6 Thermoelasticity energy harvesting

This process of energy conversion involves, further down the chain, a thermally generated mechanical transduction into electrical energy through mechanisms such as electrostatic transducers. In so doing, the system includes an intermediate step that utilizes the mechanical motion developed as a result of heat fluctuations to actuate this transduction process for a more efficient and effective electrical generation process. Electrostatic transducers, in particular, represent a promising approach due to their scalability, compatibility with microfabrication techniques, and wide-frequency operability, making them ideal for integration into advanced thermomagnetic or thermoelectric energy systems.

Electrostatic transducers effectively convert mechanical energy into electrical energy, as the mechanical input serves to switch the capacitance of the generator between two extreme values, usually a high and a low value. In most cases, this is achieved by either increasing the distance between the plates of the capacitor or reducing the area of overlap between them, thereby changing its capacitance. Electrostatic transducers can operate in two distinct modes:.

The limited charge mode keeps a constant charge and electric field on the capacitor plates. While capacitance reduces, voltage increases. In contrast, the second mode, called the limited voltage mode, uses a constant voltage across the plates. In this way, when capacitance is reduced, the charge is transferred from the capacitor to an energy storage device, allowing usable electrical energy to be produced [8,16].

Both techniques function on the basis of transforming mechanical kinetic energy into electrical energy by exerting force against the electrostatic forces acting between the capacitor plates. To facilitate this conversion, a control circuit [7,8] is often used, which is responsible for biasing the transducer and extracting the generated energy. The design and layout of the control and biasing circuitry are tightly linked to the transducer's specific operational mode. Examples of such circuits can be seen in [8,16].

An alternative to traditional biasing techniques is the use of an electret material, that is, a material with a permanently implanted charge within a dielectric layer. The electret eliminates the need for external biasing, so the system architecture is easier to install and may also be more reliable [8,14]. A complete comparison between the two modes of operation is made in Figure 4.3, where the differences and their uses are emphasized.

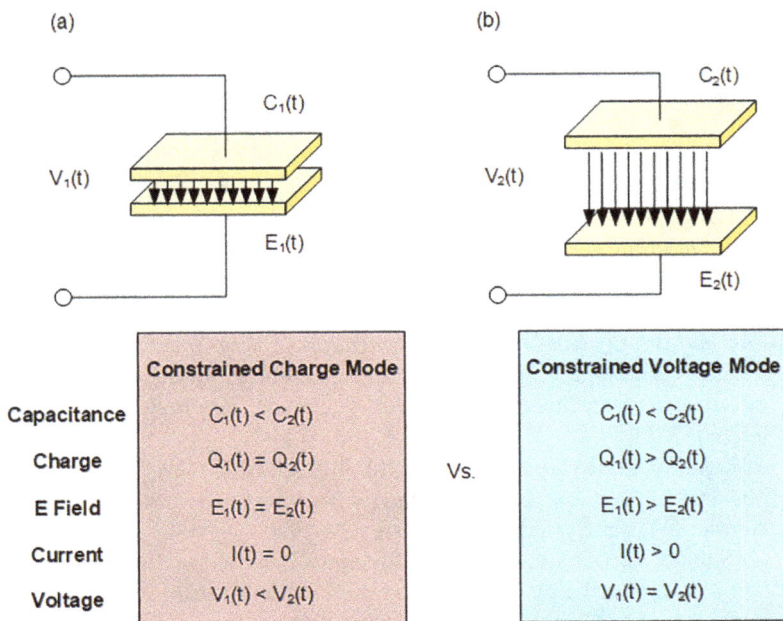

	Constrained Charge Mode		Constrained Voltage Mode
Capacitance	$C_1(t) < C_2(t)$		$C_1(t) < C_2(t)$
Charge	$Q_1(t) = Q_2(t)$	Vs.	$Q_1(t) > Q_2(t)$
E Field	$E_1(t) = E_2(t)$		$E_1(t) > E_2(t)$
Current	$I(t) = 0$		$I(t) > 0$
Voltage	$V_1(t) < V_2(t)$		$V_1(t) = V_2(t)$

Figure 4.3   A schematic of an electrostatic transducer with two states of (a) low energy and (b) high energy, as well as two modes [8]

Electrostatic transducers can effectively convert mechanical energy to electrical energy with a high potential to be utilized in compact, lightweight, and scalable energy harvesting systems. Advancing the field of material science and improving control methodologies may eventually yield even more efficient and adaptable transducer designs.

References

[1]   A. Cadei, A. Dionisi, E. Sardini, and M. Serpelloni. Kinetic and thermal energy harvesters for implantable medical devices and biomedical autonomous sensors. *Measurement Science and Technology*, 25(1):012003, 2013.

[2]   A. Cuadras, M. Gasulla, and V. Ferrari. Thermal energy harvesting through pyroelectricity. *Sensors and Actuators A: Physical*, 158(1):132–139, 2010.

[3]   A. Hossain and M.H. Rashid. Pyroelectric detectors and their applications. *IEEE Transactions on Industry Applications*, 27(5):824–829, 1991.

[4]   C.-J. Hsu, S.M. Sandoval, K.P. Wetzlar, and G.P. Carman. Thermomagnetic conversion efficiencies for ferromagnetic materials. *Journal of Applied Physics*, 110(12), 2011.

[5]   S.R. Hunter, N.V. Lavrik, S. Mostafa, S. Rajic, and P.G. Datskos. Review of pyroelectric thermal energy harvesting and new MEMS-based resonant energy conversion techniques. In *Energy Harvesting and Storage: Materials, Devices, and Applications III*, volume 8377, page 83770D. International Society for Optics and Photonics, 2012.

[6]   C. Knight and J. Davidson. Thermoelectric energy harvesting as a wireless sensor node power source. In *Active and Passive Smart Structures and Integrated Systems 2010*, volume 7643, pages 417–427. SPIE, 2010.

[7]   S. Meninger, J.O. Mur-Miranda, R. Amirtharajah, A. Chandrakasan, and J. Lang. Vibration-to-electric energy conversion. In *Proceedings of the 1999 International Symposium on Low Power Electronics and Design*, pages 48–53, 1999.

[8]   S. Percy, C. Knight, S. McGarry, A. Post, T. Moore, and K. Cavanagh. *Thermal Energy Harvesting for Application at MEMS Scale*. Springer, 2014.

[9]   S.K.T. Ravindran, T. Huesgen, M. Kroener, and P. Woias. A self-sustaining micro thermomechanic-pyroelectric generator. *Applied Physics Letters*, 99(10), 2011.

[10] M.I. Rudnicki, R.H. Chesworth, and L.T. Harmison. Design of a sup238 Pu-powered implantable thermal engine for heart assist devices. Technical report, Aerojet Nuclear Systems Co., Azusa, CA, 1970.

[11] G. Sebald, E. Lefeuvre, and D. Guyomar. Pyroelectric energy conversion: Optimization principles. *IEEE Transactions on Ultrasonics, Ferroelectrics, and Frequency Control*, 55(3):538–551, 2008.

[12] S. Sedky, A. Kamal, M. Yomn, *et al.* Bi2Te3 as an active material for MEMS based devices fabricated at room temperature. In *Transducers 2009–2009*

International Solid-State Sensors, Actuators and Microsystems Conference, pages 1035–1038. IEEE, 2009.

[13] M. Shen, W. Li, M.-Y. Li, *et al.* High room-temperature pyroelectric property in lead-free BNT-BZT ferroelectric ceramics for thermal energy harvesting. *Journal of the European Ceramic Society*, 39(5):1810–1818, 2019.

[14] T. Sterken, P. Fiorini, K. Baert, R. Puers, and G. Borghs. An electret-based electrostatic μ-generator. In *Transducers'03. 12th International Conference on Solid-State Sensors, Actuators and Microsystems. Digest of Technical Papers (Cat. No. 03TH8664)*, volume 2, pages 1291–1294. IEEE, 2003.

[15] J.W. Stevens. Optimal placement depth for air–ground heat transfer systems. *Applied Thermal Engineering*, 24(2–3):149–157, 2004.

[16] E.O. Torres and G.A. Rincón-Mora. A 0.7-μm BiCMOS electrostatic energy-harvesting system IC. *IEEE Journal of Solid-State Circuits*, 45(2):483–496, 2010.

[17] R. Venkatasubramanian, E. Siivola, T. Colpitts, and B. O'quinn. Thin-film thermoelectric devices with high room-temperature figures of merit. *Nature*, 413(6856):597–602, 2001.

[18] R.W. Whatmore. Pyroelectric devices and materials. *Reports on Progress in Physics*, 49(12):1335, 1986.

[19] J. Xie, X.P. Mane, C.W. Green, K.M. Mossi, and K.K. Leang. Performance of thin piezoelectric materials for pyroelectric energy harvesting. *Journal of Intelligent Material Systems and Structures*, 21(3):243–249, 2010.

[20] E.-J. Yoon, J.-T. Park, and C.-G. Yu. Thermal energy harvesting circuit with maximum power point tracking control for self-powered sensor node applications. *Frontiers of Information Technology & Electronic Engineering*, 19(2):285–296, 2018.

Chapter 5
Biofuel energy harvesting in implantable applications

Biofuel cells (BCs) use living organisms to generate electricity and were first demonstrated in implantable applications in 1974 [20]. The BC schematic is shown in Figure 5.1(a), which shows that BCs use biocatalysts to produce power. These cells can convert biochemical energy to electricity via biochemical reactions. Living organisms in the human body have biofuels (e.g., glucose in blood) that are capable of generating power in the microwatt range [1]. The biofuel is oxidized at the anode of the BCs, where electrons are released, and oxygen is reduced at the cathode. The catalyst in BCs can be an enzyme that can directly convert the chemical energy of carbohydrates to electric energy [1].

5.1 Principles of biofuel energy harvesting

Biochemical energy harvesting is a new approach to powering gadgets that utilizes the chemical energy within the human body. This technology takes advantage of the natural abundance of bioavailable chemicals like glucose to generate electrical energy through BCs. Unlike common batteries, which have finite chemical reserves and must be replaced or recharged regularly, BCs tap into the continuous supply of fuel provided by the metabolism of the host. In this way, the life of an organism can, in principle, be infinitely long, provided that its metabolic function remains stable [27].

The human body has an enormous reservoir of chemical energy more than 100 watts of potential power stored in organic substances such as glucose, lactate, and oxygen [11]. The comparison from biofuel harvesting and the other energy harvesting is shown in Figure 5.2. BCs are specifically designed to utilize this energy. BCs use biocatalysts, such as enzymes or bacteria, to convert these chemicals into electrical energy through oxidation–reduction reactions. For fuel, for example, glucose is the most commonly used in BCs, is highly abundant in blood and other body fluids, and therefore offers an excellent and renewable source of energy.

5.2 Materials and methods

Mesoporous carbons are used for the anodic biocatalyst for glucose oxidation [1]. In 2009, M. Zhao *et al.* showed a comparison between the electrical performance of

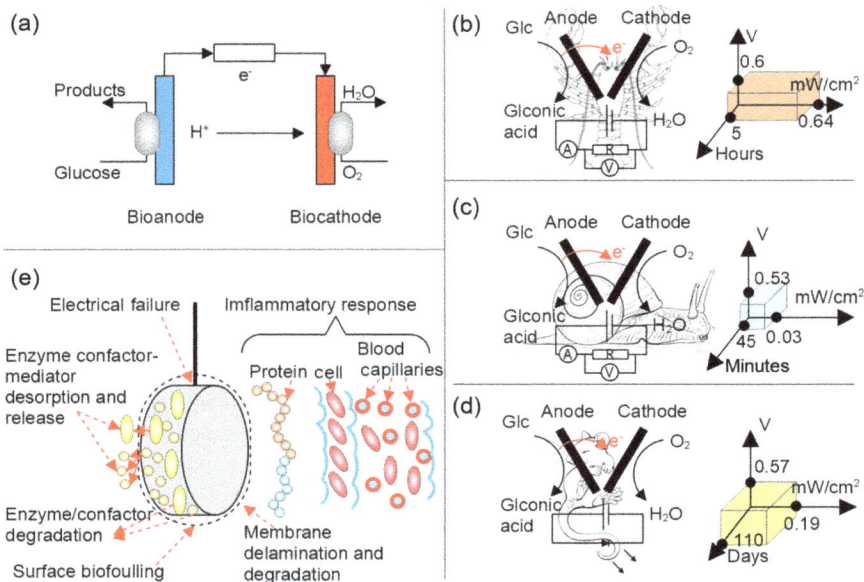

Figure 5.1   Schematic diagram and mechanism of the biofuel cell (BC). (a) shows the schematic of the BC. (b) shows the several potential sources of malfunction for bioelectrodes after implantation. (c) shows the implantable BC in a living lobster with output voltage, output power density, and implanting period [15]. (d) shows the implantable BC in a living snail with output voltage, output power density, and implanting period [8]. (e) shows the implantable BC in a living rat with output voltage, output power density, and implanting period [28]

the BCs based on the mesoporous carbons and carbon nanotubes (CNTs) [30]. The electric performance of mesoporous carbons based structure was much more impressive than CNT-based structure, which are ($V_{oc} = 0.82$ V and $P_{max} = 38.7$ μW/cm^2) and ($V_{oc} = 0.75$ V and $P_{max} = 2.1$ μW/cm^2), respectively. Thus, it is a strong proof that mesoporous carbons will be the novel type of robust and advanced material for electrode [1,30]. The BC is modified by DNA wrapped single-walled carbon nanotubes (SWNTs). It was found that the electric characteristics increased attributing to the immobilization of glucose oxidase (GOx) and active site protection. Moreover, this novel structure will provide a tremendous power density of 730 to 760 μW/cm^2 with a duration of a week and the stability of BCs will be improved [12]. A dual power harvester combined with a thermal electric generator (TEG) and BCs was embedded with 0.18 μW complementary metal-oxide-semiconductor integrated circuit. Simultaneous maximum power extraction of two power generators is led to a control circuit to improve efficiency and diminish the switching loss. Both harvesters obtain remarkable electric power, which is 23 and 29 μW in TEG, respectively [9]. The BC was

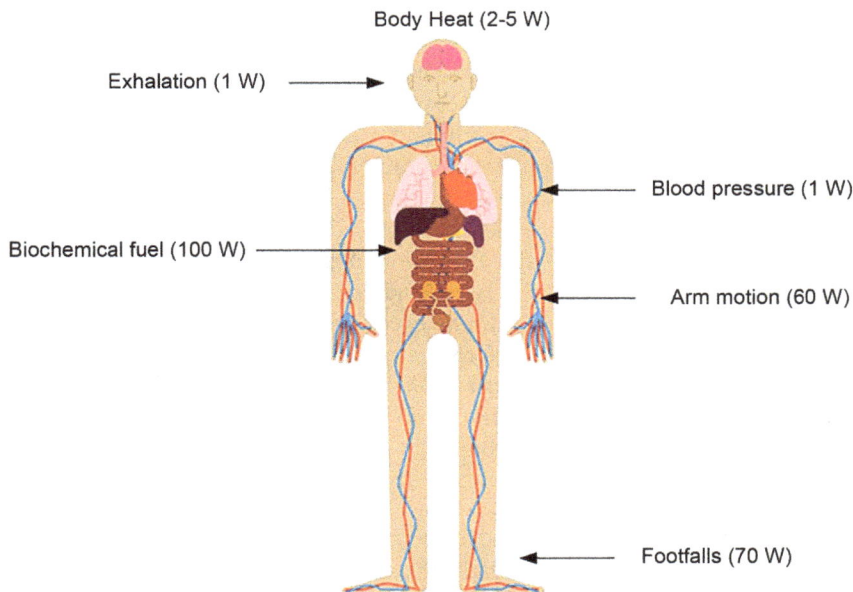

Figure 5.2   The energy inside a human body. Recreate from [27]

fabricated with Lac-GAfCs-MWCNTs/GC cathode and GOx-GAfCS-MWCNTs/GC anode, respectively, and the cell operates inside the membrane-separated acetate solution at pH 5. The output voltage, current density, and power density are 0.19 V, 114 $\mu A/cm^2$, and 9.6 $\mu W/cm^2$, respectively [22]. The BC has been implanted inside distinctive living organism such as lobster [15], snail [8], claim [21], rat [2,28], and rabbit [17]. The dehydrogenase-laccase (Anode-Cathode) structure is applied to implant in the lobster [15] shown in Figure 5.1(b), while the schematic of the BC implanted in snail [8] and rat [28] are shown in Figure 5.1(c) and (d), respectively.

Among the issues with biochemical energy harvesting is the low-power conversion efficiency, which means that a large area is required. This makes them infeasible for implantable power harvesting applications. Meanwhile, there are a lot of drawbacks that have an negative effect after implanting, which is shown in Figure 5.1(e).

5.3   Enzymatic biofuel cell

In the past five decades, enzymatic BCs have achieved incremental yet significant improvements. The concept originally emerged in the 1960s with the finding that oxidoreductase enzymes, more commonly referred to as redox enzymes, were able to catalyze fuel oxidation at the anode of a BC. These early devices, however, had very low current densities and were unsuitable for most applications [26,27].

In fact, until the late 1990s and early 2000s, technological advancements allowed researchers to develop prototypes that could achieve current densities close to mA/cm^2 in a lab setting [4,18,27]. One of the more ambitious uses that resulted from this pivotal moment was the creation of implantable BCs. However, early attempts to in vivo applications faced many problems, especially for long-term stability and reliable performance, and delayed further progress for several decades.

In 2003, Heller *et al.* [16,27] performed a seminal experiment to demonstrate the implantation of a glucose enzymatic biofuel cell (GBC) in a grape. This achievement caused a new wave of research in implantable devices and re-awakened interest in BC technology. By the 2010s, the field had reached a level of maturity at which researchers could experiment with BCs in living organisms. In 2010, for example, a GBC was implanted in a rat [5,27]. These studies showed that enzymatic BCs could now be used in vivo to produce electricity over an extended period of time.

Despite these advances, several challenges are still crucial for the successful realization of enzymatic BCs for implantable applications. The three main obstacles include the following: Enzyme stability and yield: Enzymes used in BCs degrade over time, which diminishes their activity and overall lifetime. Biocompatibility: All components of the BC must be biocompatible to prevent immunological reactions or other adverse effects. Biofouling is the deposition of living molecules on the surface of a BC that impede performance and diminish efficiency.

These subjects have been the focus of significant research. El-Ichi *et al.* [6,27] tackled the issues of biocompatibility and stability by fabricating a 3D nanofibrous biocathode from compressed chitosan cross-linked with genipin and enhanced with CNTs and the enzyme laccase. For better biocompatibility, the biocathode was modified by forming a thin layer of chitosan cross-linked with genipin. This distinctive construction gave good performance in an in vitro environment: running at physiological pH (7.4), ionic strength (140 mM NaCl), it is capable of supplying -0.3 mA/ml continuous discharging over a period of 20 days. Even more impressively, when implanted in a rat, the biocathode continued to operate for 167 days, indicating significant advances toward long-term stability and biocompatibility.

In the future, the advancement of enzymatic BCs depends on overcoming remaining technical hurdles through interdisciplinary innovation. The immobilization of enzymes on nanostructured surfaces or the use of modified enzymes with enhanced degradation resistance are only a few ways to enhance enzyme stability. Methods for mitigating biofouling, such as improved surface coatings or anti-fouling treatments, may further improve long-term performance.

In just the past decade, enzymatic BCs have evolved from an imaginary vision to something quite realistic for implantable power sources. Enzymatic BCs could revolutionize medical implant technology by offering them miniaturized, fully biocompatible, self-sustained power due to the recent material advances with concomitant advances in biochemistry and devices' design.

5.4   Conventional biofuel cell and microbial biofuel cell

Conventional BCs use two types of electron transfer: Direct electron transfer (DET) and mediated electron transfer (MET). Among these, glucose/O$_2$-based DET systems typically have lower current and power densities than MET-based counterparts [19].

This variance is mainly due to the specific anodic bioelements used in DET enzymatic anodes, such as cellobiose and glucose dehydrogenases [7,19]. GOx, a highly active and selective oxidoreductase commonly used in MET-based glucose oxidizing bioanodes [14], these redox enzymes have comparatively low catalytic activity toward glucose.

Despite these constraints, there have been notable exceptions to the creation of DET-based bioanodes. For example, novel mediator-free bioanodes based on mechanically compressed CNT disks with integrated GOx have shown substantial activity [1]. However, most MET-based BCs outperform the power outputs of DET-based devices. It is illogical to consider that, in theory, DET-based BCs should be able to give higher operation voltages. In fact, this theoretical advantage is the lack of mediators. In MET systems, they cause voltage loss related to the potential difference between the enzyme's active site and the mediator. In addition to the electrical efficiency considerations, the use of mediators in MET systems has further potential issues for implantable applications. Mediators are often composed of chemicals that are toxic or unstable under physiological conditions, increasing the potential for harmful effects upon use in vivo. DET-based systems, by definition, do not require the use of mediators and therefore have reduced risks, making them more suitable for implantable applications [27].

Recent advances in electronic technologies, especially the shrinkage of components down to micro- and nanoscales, have radically changed the power requirements for many systems. Modern implantable devices usually need only microwatts of power, while even low-output BCs are able to efficiently meet such energy demands. DET-based BCs hold a special place for such applications due to their simpler architecture and lower biocompatibility issues. Moreover, ongoing advances in materials science on the design of nanostructured electrodes and highly conductive biocompatible materials may increase the efficiency of DET-based devices. These developments can help to close the gap in performance between DET and MET systems, rendering DET-based BCs more competitive while still retaining their intrinsic biocompatibility and safety advantages.

Microbial BCs are among the first BC technologies, developed as an alternative to conventional fuel cells in a sustainable manner [11]. The fundamental ideas and performance characteristics of several microbial BCs have been well covered in the scientific literature over the years [10]. Among the several techniques for bioenergy conversion, microbial fuel cells (MFCs) stand out as a very fuel-efficient way to use complex organic substances such as carbohydrates. The principle behind the oxidation reaction in MFCs involves a sequence of enzyme-catalyzed reactions with the aid of microorganisms, thereby allowing energy extraction from complex substrates [13].

The operating principles of MFCs can be defined using two classic methods: (1) Production of electroactive metabolites: Microorganisms ferment fuel substrates to synthesize electroactive chemicals. Such metabolites serve as shuttles to carry electrons to the electrode surface [29]. (2) Redox mediators: In this method, redox-active chemicals are used for the transport of electrons from microbial metabolic pathways to electrodes, improving electron transfer efficiency [27].

Recent improvements have increased MFC uses and efficiency. A pioneering MFC that harnesses energy from the marine sediment–seawater interface was reported recently [24]. This system utilizes inherent electrochemical gradients in maritime conditions for sustainable remote energy production.

Furthermore, the discovery of hybrid BCs that combine microbial and enzymatic systems has opened up new possibilities for photosynthesis energy conversion [23]. This innovative photosynthetic BC combines the benefits of microbial metabolic processes with enzymatic catalysis to create a more adaptable platform for bioenergy applications.

Unconventional biomass-fueled ceramic fuel cells have been realized, proving the versatility of BC technology to a variety of substrates and environmental circumstances [11]. The concept of "gastrobots" – hybrid robots fueled by MFCs – is an exciting use of MFCs in robotics [25]. These robots are powered by energy produced by microbial processes, demonstrating the potential of BCs to power autonomous systems in a sustainable manner. Another major achievement is the development of mediatorless MFCs capable of directly oxidizing glucose. Such devices have achieved current densities up to 3 $\mu A/cm^2$, although at an unknown cell voltage [3]. The absence of mediators in these systems simplifies system architecture, reduces costs, and enhances biocompatibility, making them suitable for use as implantable medical devices.

In applications related to implantable devices, traditional BCs are far more appropriate and advanced compared to microbial BCs. Traditional BCs are designed for compact operation within the physiological milieu, deriving energy from easily available body fuels such as glucose or lactate. These systems, and MET in particular, enable significantly higher power densities, compatible with the operating energies of implantable pacemakers and glucose monitors. Whereas the addition of mediators to MET systems complicates this technology, the architecture for MET systems is far less cumbersome and more viable than what is required for in vivo devices. However, the potential toxicity of mediators and limited stability of enzymes in these systems create barriers, whereas the advances in enzyme immobilization and biocompatible materials are easing the concerns.

In contrast, MFCs possess some serious drawbacks that question their feasibility for implantable applications. These systems require the action of living microbes for the degradation of organic substrates, which makes it more challenging to maintain the systems inside the human body. In this controlled physiological environment, their use is impeded by microbial survival, immune system responses, and risk of infection. Besides this, MFCs also usually give a very low power density, which may be insufficient to cover the energy demand of state-of-the-art implantable devices. While MFCs find the best application in long operating and self-replenishing fuel sources, like environmental energy harvesting, their dependence on microbial activity and possibility for overgrowth of biofilm limits their biomedical implant applications. Even if the MFC is less advantageous compared with a conventional BC, large strides in lab-on-a-chip and microelectromechanical systems have been made possible by the effective shrinking of sensors, actuators, and systems.

References

[1] A.A. Babadi, S. Bagheri, and S.B.A. Hamid. Progress on implantable biofuel cell: Nano-carbon functionalization for enzyme immobilization enhancement. *Biosensors and Bioelectronics*, 79:850–860, 2016.

[2] J.A. Castorena-Gonzalez, C. Foote, K. MacVittie, *et al.* Biofuel cell operating in vivo in rat. *Electroanalysis*, 25(7):1579–1584, 2013.

[3] S.K. Chaudhuri and D.R. Lovley. Electricity generation by direct oxidation of glucose in mediatorless microbial fuel cells. *Nature Biotechnology*, 21(10):1229–1232, 2003.

[4] T. Chen, S.C. Barton, G. Binyamin, *et al.* A miniature biofuel cell. *Journal of the American Chemical Society*, 123(35):8630–8631, 2001.

[5] P. Cinquin, C. Gondran, F. Giroud, *et al.* A glucose biofuel cell implanted in rats. *PLoS One*, 5(5):e10476, 2010.

[6] S. El Ichi, A. Zebda, J.-P. Alcaraz, *et al.* Bioelectrodes modified with chitosan for long-term energy supply from the body. *Energy & Environmental Science*, 8(3):1017–1026, 2015.

[7] M. Falk, V. Andoralov, Z. Blum, *et al.* Biofuel cell as a power source for electronic contact lenses. *Biosensors and Bioelectronics*, 37(1):38–45, 2012.

[8] L. Halámková, J. Halámek, V. Bocharova, A. Szczupak, L. Alfonta, and E. Katz. Implanted biofuel cell operating in a living snail. *Journal of the American Chemical Society*, 134(11):5040–5043, 2012.

[9] J. Katic, S. Rodriguez, and A. Rusu. A high-efficiency energy harvesting interface for implanted biofuel cell and thermal harvesters. *IEEE Transactions on Power Electronics*, 33(5):4125–4134, 2017.

[10] E. Katz, A.N. Shipway, and I. Willner. Biochemical fuel cells. *Handbook of Fuel Cells – Fundamentals, Technology and Applications*, 1:355–381, 2003.

[11] E. Katz and K. MacVittie. Implanted biofuel cells operating in vivo methods, applications and perspectives. *Energy & Environmental Science*, 6(10): 2791–2803, 2013.

[12] J.Y. Lee, H.Y. Shin, S.W. Kang, C. Park, and S.W. Kim. Improvement of electrical properties via glucose oxidase-immobilization by actively turning over glucose for an enzyme-based biofuel cell modified with DNA-wrapped single walled nanotubes. *Biosensors and Bioelectronics*, 26(5):2685–2688, 2011.

[13] D. Leech, P. Kavanagh, and W. Schuhmann. Enzymatic fuel cells: Recent progress. *Electrochimica Acta*, 84:223–234, 2012.

[14] R. Ludwig, W. Harreither, F. Tasca, and L. Gorton. Cellobiose dehydrogenase: A versatile catalyst for electrochemical applications. *ChemPhysChem*, 11(13):2674–2697, 2010.

[15] K. MacVittie, J. Halámek, L. Halámková, *et al.* From "cyborg" lobsters to a pacemaker powered by implantable biofuel cells. *Energy & Environmental Science*, 6(1):81–86, 2013.

[16] N. Mano, F. Mao, and A. Heller. Characteristics of a miniature compartmentless glucose–O_2 biofuel cell and its operation in a living plant. *Journal of the American Chemical Society*, 125(21):6588–6594, 2003.

[17] T. Miyake, K. Haneda, N. Nagai, *et al.* Enzymatic biofuel cells designed for direct power generation from biofluids in living organisms. *Energy & Environmental Science*, 4(12):5008–5012, 2011.

[18] G.T.R. Palmore, H. Bertschy, S.H. Bergens, and G.M. Whitesides. A methanol/dioxygen biofuel cell that uses Nad+-dependent dehydrogenases as catalysts: Application of an electro-enzymatic method to regenerate nicotinamide adenine dinucleotide at low overpotentials. *Journal of Electroanalytical Chemistry*, 443(1):155–161, 1998.

[19] D. Pankratov, Z. Blum, D. Suyatin, V. Popov, and S. Shleev. Self-charging electrochemical biocapacitor. *ChemElectroChem*, 1(2):343–346, 2014.

[20] J.R. Rao, G. Richter, F. Von Sturm, E. Weidlich, and M. Wenzel. Metaloxygen and glucose-oxygen cells for implantable devices. *Biomedical engineering*, 9(3):98, 1974.

[21] A. Szczupak, J. Halámek, L. Halámková, V. Bocharova, L. Alfonta, and E. Katz. Living battery–biofuel cells operating in vivo in clams. *Energy & Environmental Science*, 5(10):8891–8895, 2012.

[22] Y. Tan, W. Deng, B. Ge, Q. Xie, J. Huang, and S. Yao. Biofuel cell and phenolic biosensor based on acid-resistant laccase–glutaraldehyde functionalized chitosan–multiwalled carbon nanotubes nanocomposite film. *Biosensors and Bioelectronics*, 24(7):2225–2231, 2009.

[23] S. Tsujimura, A. Wadano, K. Kano, and T. Ikeda. Photosynthetic bioelectrochemical cell utilizing cyanobacteria and water-generating oxidase. *Enzyme and Microbial Technology*, 29(4–5):225–231, 2001.

[24] X. Wei and J. Liu. Power sources and electrical recharging strategies for implantable medical devices. *Frontiers of Energy and Power Engineering in China*, 2:1–13, 2008.

[25] S. Wilkinson. "Gastrobots" – Benefits and challenges of microbial fuel cells in foodpowered robot applications. *Autonomous Robots*, 9:99–111, 2000.

[26] A.T. Yahiro, S.M. Lee, and D.O. Kimble. Bioelectrochemistry: I. Enzyme utilizing bio-fuel cell studies. *Biochimica et Biophysica Acta (BBA)-Specialized Section on Biophysical Subjects*, 88(2):375–383, 1964.

[27] A. Zebda, J.-P. Alcaraz, P. Vadgama, *et al.* Challenges for successful implantation of biofuel cells. *Bioelectrochemistry*, 124:57–72, 2018.

[28] A. Zebda, S. Cosnier, J.-P. Alcaraz, *et al.* Single glucose biofuel cells implanted in rats power electronic devices. *Scientific Reports*, 3(1):1–5, 2013.

[29] Q. Zhong, J. Yan, X. Qian, T. Zhang, Z. Zhang, and A. Li. Atomic layer deposition enhanced grafting of phosphorylcholine on stainless steel for intravascular stents. *Colloids and Surfaces B: Biointerfaces*, 121:238–247, 2014.

[30] M. Zhou, L. Deng, D. Wen, L. Shang, L. Jin, and S. Dong. Highly ordered mesoporous carbons-based glucose/O_2 biofuel cell. *Biosensors and Bioelectronics*, 24(9):2904–2908, 2009.

Chapter 6

RF energy harvesting and wireless power transfer in implantable applications

6.1 Principles of RF energy and wireless power transfer (WPT) scheme

Radio frequency (RF) energy harvesting involves scavenging power from electromagnetic radiation. The first RF energy harvesters were demonstrated in pacemakers back in 1969 [9]. RF waves can be classified as near field (NF) or far field (FF), depending on the electromagnetic waves in the different distance (specified by Fraunhofer distance) [23]. The distribution of the electrical field is shown in Figure 6.1(a), and blocks as well as design aspects are shown in Figure 6.1(b).

Near-field RF power harvesting can be described using the following equations [21]:

$$P_{in} = \frac{M}{L_{f1}R_2} \left(\frac{M}{L_f}V_1 - V_2 \right) V_1 \tag{6.1}$$

$$P_{out} = \frac{1}{R_2} \left(\frac{M}{L_f}V_1 - V_2 \right) V_2 \tag{6.2}$$

where M is the mutual inductance between the transmitting and receiving coils, L_f is the inductance of the transmitting coil at frequency f, V_1 is the supply voltage at the transmitting end, V_2 is the charging voltage at the receiving end, R_2 is the resistance at the receiving end. The far-field RF power harvesting can be described as the following equation [23]:

$$P_{out} = \frac{P_T G_T G_R \lambda^2}{(4\pi R)^2} \tag{6.3}$$

where P_T is the power in the transmitting antenna, G_T and G_R are the gain of the transmitter and receiver antennas, respectively. λ is the wavelength of the electromagnetic wave. The schematic of the integrate coils of all the materials is shown in Figure 6.1(c) [24], and the coil is embedded in the rectifier shown in Figure 6.1(d) and (e).

When using RF signals for energy transfer, one must distinguish between nonradiative and radiative approaches. Nonradiative RF energy transfer is based on

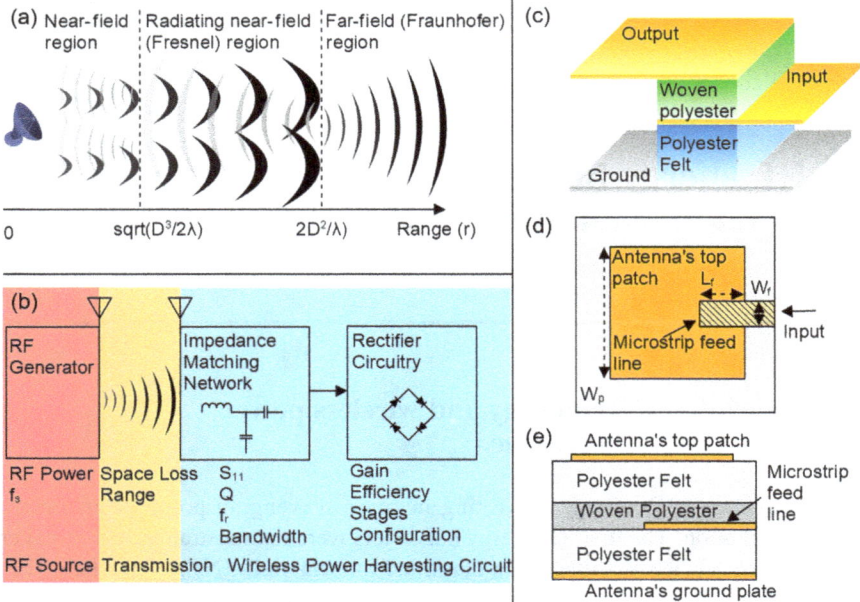

Figure 6.1   Schematic diagram and mechanism of the radio frequency (RF) generator. (a) shows distribution of NF and FF regions in space [24]. (b) shows details of the conceptual block diagram of an RF power harvesting system [24]. (c)–(e) show structure of all-fabric antenna and layout of the rectifier on Duroid-5880 (35 mm×23 mm) including the fabric-to-printed circuit board (PCB) coupling interface (top view is to scale) [1]

inductive coupling, where energy is transferred over short distances by means of magnetic fields. In contrast, in radiative RF energy transfer, radio waves are transmitted and received to deliver energy over longer distances. Furthermore, a clear differentiation must be made between RF energy harvesting and RF energy transfer. RF energy harvesting involves the collection of energy from surrounding RF signals. These are usually transmitted for telecommunication purposes, such as radio and television broadcasting, global system for mobile communication (GSM), and Wi-Fi. In such cases, the energy captured is a byproduct of transmissions that were not originally meant for power delivery. However, RF energy transfer involves the use of an independent RF source with its operation and design aimed at wirelessly powering a device. The signals transmitted in RF energy transfer are intentional and directed, unlike the incident signals utilized in energy harvesting [5].

Strictly speaking, RF energy harvesting can be considered a subset of RF energy transfer, since both involve the conversion of RF signals into electrical energy that can be used. However, the defining distinction lies in the source and intent of the transmitted signals: In RF energy harvesting, the source is ambient and unintentional, while in RF energy transfer, the source is deliberate and engineered for power delivery. This differentiation is very important in the design and optimization of energy systems, as the requirements for ambient energy harvesting, such as low-power operation and the ability to capture widely dispersed signals, differ significantly from those of dedicated RF energy transfer that typically aims for higher efficiency and directed power transmission. Both methods hold promise for advancing wireless energy technologies, with applications ranging from self-sustaining IoT devices to remotely powered sensors and implants [5].

6.2 Structure and method

A neural interface microsystem was powered by an inductive link in 2 MHz that was supplied by a Li-ion battery [3,4]. The power link consisted of a 210-pF ceramic capacitor and a 27-turn 27-mm AWG 40 strand Litz wire with inductance of 2 μH. This coil was advantageous in terms of a high quality factor (Q) of 75.4 with 2 MHz [3]. A copper based on the ceramic coil was embedded in a 9-mm^2 chip that is implanted in the human retina to produce artificial vision. Patients tested five subjects such as light perception, localization, and motion detection. The copper in the polyimide coil was applied at the FF frequency of 910 MHz was operated. In this case, the possibility of RF energy harvesting within moving small animals was approved and completely wireless behavior can be controlled [3,12]. Energy was coupled from a copper power link in Ferro solution to resonant energy into a head-mounted device with a maximum magnetic field of 300 A/m. This was implanted in vivo in a rat inside a cage-shaped power transducer at a resonant frequency of 120 kHz. The device was orthogonal to the cage to optimize magnetic coupling [3,26]. In 2010, two on-chip antennas were embedded in the circuit using 0.18 μm technology. A remarkable power scavenging distance of 7.5 cm was achieved and, in the meantime, the down-link harvested the 13.2 μW/cm^2 power. There was no-off chip components used in this research [2]. In 2011, a loop antenna was applied on an intraocular monitor, and the biocompatible methacrylate plastic and small size make it a good candidate in the application of retinas. The power density of 28.33 μW/cm^2 was achieved in 1.5 cm at −10.5-dBm RF sensitivity [19]. In 2018, circular polarization was considered in antenna design, and the antenna had a relatively smaller size and can obtain better electromagnetic radiation. After being tested on 4 mm skin, 9.65 μW/cm^2 power density was obtained at a distance of 40 cm at 915 MHz [14]. It should be emphasized that the power density of near-field RF is unpredictable compared to far-field RF, which is normally considered uniform [23]. The different magnetic flux densities in the NF and FF ranges make the RF harvester face different challenges associated with these ranges. Near-field RF requires perfect impedance matching and antenna alignment, which is still a challenge for designers. According to the low FF magnetic flux density, it is difficult to harvest sufficient energy to start up the circuit by using far-field RF harvester, where the start-up thresholds are different in different CMOS technologies [16]. In addition, the frequency of far-field RF harvester is limited up to GHz range as a result of tissue loss [16].

6.3   Near-field nonradiative coupling

Nonradiative RF energy transfer is generally used in inductive coupling systems in which two coils are close together to ease energy transmission. Depending on the distance and range, it consists of an NF coupling and a mid-field coupling. This principle is the foundation for innovations such as the Qi wireless charging standard [2]. When two coils are brought close together, their magnetic fields combine to create an electrical transformer. By tuning the circuit with capacitors, a resonant coupling system is established, dramatically improving power transfer efficiency (PTE) [5,13].

Figure 6.2 shows a resonant coupling setup in which an RF source is coupled to a tuned transmitting coil (L_s) with a capacitor C_s that creates the resonance. It also shows a receiving coil L_r that is in close proximity, connected to a C_r capacitor C_r. The energy delivered to the receiving coil is rectified and then regulated before being stored or used to power a load indicated by the resistance R_L. Losses in the transmitting and receiving circuits are represented by the R and r resistances, respectively.

The cardinal benefit of resonant inductive coupling is the high PTE in short ranges. Resonant tuning ensures that energy transfer is highly effective within a narrow frequency range, thus enabling the surrounding devices to be powered with much more efficiency. This makes the technology excellent for applications like wireless charging pads and implantable medical devices (IMDs), where the receiver is within proximity to the transmitter. In addition, adherence to developed standards, such as Qi, allows compatibility and universality of function within consumer electronics [15].

Despite its high efficiency, resonant inductive coupling has considerable limitations, most noticeably regarding the operating range. The efficiency of energy transfer drops off very sharply when the distance between the transmitting and receiving coils exceeds about one coil diameter. This dramatic drop in performance limits its applicability to short-range applications, making it unsuitable for systems that require longer distance power transfer. Furthermore, the proximity requirement requires the devices to remain immobile and perfectly aligned during the charging process, reducing flexibility for applications that require mobility or movement [5,15].

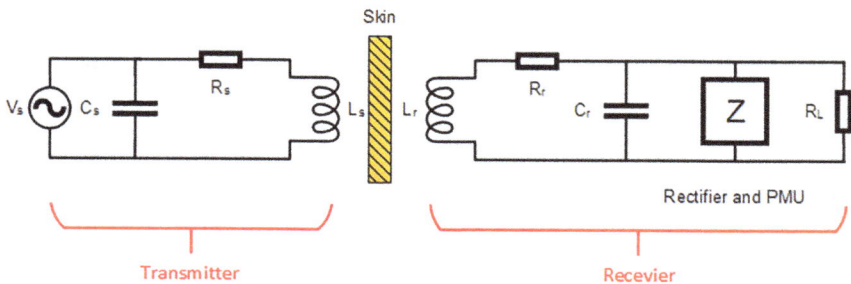

Figure 6.2   Schematic diagram and mechanism of the resonant coupling (nonradiative) RF energy transfer system [5]

Another issue is that resonant inductive devices are not effective in scavenging the RF energy from ambient electromagnetic sources. The very short effective range and abrupt energy dropoff make it practically impossible to scavenge substantial power from the widely scattered RF signals provided by GSM, Wi-Fi, and other communication technologies [5,25].

The energy transfer of RF has to be of a radiative type for longer distances. Radiative transfer works in such a way that it sends electromagnetic waves out to space, which can then be received by a receiving antenna positioned further away. Unlike inductive coupling, radiative transfer does not require close alignment or proximity between the transmitter and receiver; hence, it is better suited for applications such as RF energy harvesting and remote power distribution. However, radiative transfer also has its problems, which include a reduced efficiency because of energy dispersion and the need for directional antennas that efficiently focus transmitted energy.

Although resonant inductive coupling is pretty effective over short ranges, its limitations make the use of radiative transfer techniques necessary for applications requiring longer-range power delivery or energy harvesting.

According to the coupling method, NF coupling is also classified by NF inductive coupling (NFIC) and capacitive coupling.

6.3.1 Near-field inductive coupling

NFIC is a well-studied technique in the field of RF energy harvesting, notably for implanted devices. This method uses the theory of electromagnetic induction to harvest usable energy from surrounding RF sources, making it a promising solution for powering low-energy devices like medical implants. In NFIC systems, a transmitting coil (TX) produces a time-varying magnetic field that induces an electromotive force (EMF) in a receiving coil (RX). The RX harvests the magnetic flux created by the TX and turns it into electricity for use in the implanted device.

NFIC-based RF harvesting employs magnetic coupling to transmit energy from the TX to the RX coils. The alignment and distance between the two coils, as well as the frequency and strength of the magnetic field, all have a considerable influence on the process's efficiency. However, there are certain challenges.

Several techniques can improve NFIC's performance in RF energy harvesting:

Increasing the magnetic field:

- Increasing the current in the TX coil to produce a stronger field while remaining within acceptable tissue exposure limits.
- To improve coupling efficiency, the spacing between TX and RX should be reduced.

Increasing magnetic field variability:

- Using higher operating frequencies to provide a faster-changing magnetic field, albeit this must be done cautiously to avoid tissue damage.
- Enhancing flux linkage.

To maximize the collected flux, the spatial alignment of the TX and RX coils is being optimized. Optimize coil shapes to focus the magnetic field and increase energy collection efficiency.

PTE is one of the most important factors in assessing performance in NRIC systems. PTE can be defined as the power given to the load, for example, an implanted device, versus the power drawn from the transmitting source. Achieving high PTE is difficult, especially in practical systems where coil losses such as eddy current losses and proximity effects, weak coupling due to misalignment, huge separations, transmitter–receiver mismatches, and tissue absorption drastically reduce efficiency. These losses are especially problematic for IMDs, as PTE directly impacts tissue heat dissipation and implant battery recharge cycles.

In order to enhance PTE, resonance tuning is utilized, a technique first presented by Tesla [22]. Under weak coupling conditions, the tuning of the receiving coil at the resonance frequency of the transmitting coil improves the efficiency of power transmission. Resonant tuning is very useful in NRIC systems, where it reduces energy losses while increasing overall performance.

Two major topologies are commonly used to deliver power in NRIC systems: series and parallel resonant circuits. Both variants behave very well for weak-coupling circumstances. However, series topology is often chosen because of its better performance in strong-coupling circumstances where higher PTE is obtained. In contrast, parallel topologies can handle much higher currents and lower voltages, and hence they may be really good for some biomedical applications where efficient power delivery at low currents is required [3].

The resistance of the load can be tuned to achieve maximum PTE using passive network impedance matching techniques; therefore, efficient energy delivery to the load is achieved [27,28]. The concept of load optimization introduces new design parameters that isolate the load impacts from overall link optimization, hence greater flexibility and performance.

When NRIC systems are applied to implanted devices, a number of issues arise:

- **Alignment and misalignment issues**

Implantable devices are used in dynamic environments, such as human bodies, where misalignments are unavoidable. Such misalignments reduce PTE and need to be improved with novel designs. Permanent magnets for coil alignment have been a solution long in practice; however, these magnets generate heat during MRI procedures and need to be removed before the scan [8]. Modern designs avoid this problem with self-aligning magnets that align inside the incident magnetic field, making them compatible with magnetic resonance scanners while continuing to meet safety criteria [29].

- **Resonance detuning due to motion**

Flexion, stretching, and other forms of mechanical movement detune the NRIC system, leading to a loss of PTE. These effects are reduced for more miniaturized coils featuring lower inductance values, while flexible coil designs incorporating serpentine-structured topologies or microfluidic channels help limit performance degradation [11]. Working at higher frequencies, typically within the MHz range, relaxes this bound on size while still allowing for flexibility.

- **Power and heat management**

Closed-loop power control systems can dynamically adjust the transmitted power to compensate for changes in load, thereby guaranteeing that power is delivered consistently and with high efficiency to the implant. The minimum electronics, such as basic load modulation, are used in these systems to maintain optimum PTE. Advanced approaches like operational frequency tuning and dynamic matching improve the robustness of the system by addressing power variations due to tissue absorption and coil misalignment.

- **Safety considerations**

For applications requiring high power delivery, such as functional muscle stimulation, it is necessary to adhere to the specific absorption rate (SAR) and other safety restrictions. High-excitation frequencies lead to considerable tissue heating, but it is possible to lower these effects while maintaining high PTE at lower frequencies, for example, in the kilohertz range. Magnetic material-based novel coil designs can achieve size reduction with adequate coupling for efficient energy transfer.

High-conductivity materials, such as copper, used in the construction of coils are one of the most important factors contributing to improved performance in NRIC. Although copper is more conductive and thus increases power transfer, its lack of biocompatibility prohibits its use for implanted applications. One possible solution involves encapsulating copper coils in biocompatible materials, making safe operation in tissue environments possible. This approach enables the benefits of electrical performance with copper while ensuring compliance with biological safety.

In general, for implantable applications, an NRIC system should be designed with resonance tuning, topology selection, appropriate alignment techniques, and better materials. In order to overcome the difficulties imposed by the changing human body environment, innovative designs in the form of self-aligned magnets, flexible geometries of microcoils, and dynamic power control systems become critical. As such, NRIC systems remain one of the growing dependable alternatives for wireless power delivery in biomedical implants, providing a delicate balance between safety, efficiency, and practicality.

6.3.2   Near-field capacitive coupling

The near-field capacitive coupling (NFCC) scheme is the electrical counterpart of the NRIC system. It works on electric field coupling between two pairs of conductors, one for the forward current flow and the other for the reverse direction. Displacement currents propagate without the need for a physical medium, which assists in WPT between tissue layers. Due to the high mutual impedance between the conductors, voltage stimulation across the external pair of wires, TX, would yield an incredibly low current. However, adding a second pair of conductors, RX, to complete the circuit significantly drops the mutual impedance, allowing current from the source and mirroring it through capacitive coupling to power the implant device [20].

The capacitive impedance between the TX and RX wires is a big challenge, since even small separations (in the order of millimeters) yield a high reactance. Consequently, the current from the source remains low, thereby limiting the power transfer potential. Special design changes must be implemented to make NFCC feasible for biomedical implants.

The system may be optimized as follows [3]. Increasing the Electrical Field Strength: The excitation voltage at the transmitter can be increased, subject to safety constraints on the maximum allowable electric field intensity in biological tissues. Decrease the TX-RX separation distance, though this is constrained by realistic implant dimensions. Increase the Rate of Change of the Electric Field: The operating frequency can be raised, which reduces capacitive reactance but may increase tissue loss. Maximizing Electric Field Magnitude: Increasing the area of the conductor (which is subject to constraints based on size, given implanted devices).

However, there exists a trade-off, and too much conduction current can cause tissue to lose heat and be lost, requiring thorough tuning of the system to meet performance and safety criteria.

In order to provide sufficient PTE, the NFCC operates at a higher frequency, normally in the range of several MHz. At these frequencies, tissue losses become the dominating factor due to conduction and relaxation losses. Conduction losses occur when current passes through biological tissues, whereas relaxation losses are frequency-dependent and caused by molecule polarization. At low frequencies, there are fewer polarization reversals per unit of time, which reduces relaxation losses. However, at extremely high frequencies, the molecules fail to polarize properly, resulting in reduced relaxation losses. However, an intermediate frequency range often causes the highest losses, and thus a careful selection of frequency should be considered [20].

The PTE of the NFCC system is further measured by the following relationship:

$$\eta = \frac{R_L}{R_L + R_T}(1 - |\Gamma|^2) \tag{6.4}$$

where R_L is the load resistance, R_T represents tissue and conductor losses, and Γ is the reflection coefficient. These parameters are optimized to minimize tissue loss by maximizing PTE.

The NFCC system consists of two bioencapsulated metallic patches normally for subcutaneous implantation. These constitute capacitors, having body tissue as the dielectric. Whenever there is power transmission, the conduction current leads to tissue losses and, therefore, the design should estimate these losses to not compromise efficiency or safety [10].

Figure 6.3 illustrates how an equivalent loss model with series resistors can be represented for loss of conductor and tissue.

One key characteristic of NFCC systems is that they can operate at high data rates, making them ideal for wireless telemetry applications. While their application in IMDs has just begun and several technological challenges need to be resolved, the application of NFCC systems in IMDs is still in its infancy. Power Transfer Fluctuations: The capacitance is more sensitive to the separation distance in small conductive

(a)

(b)

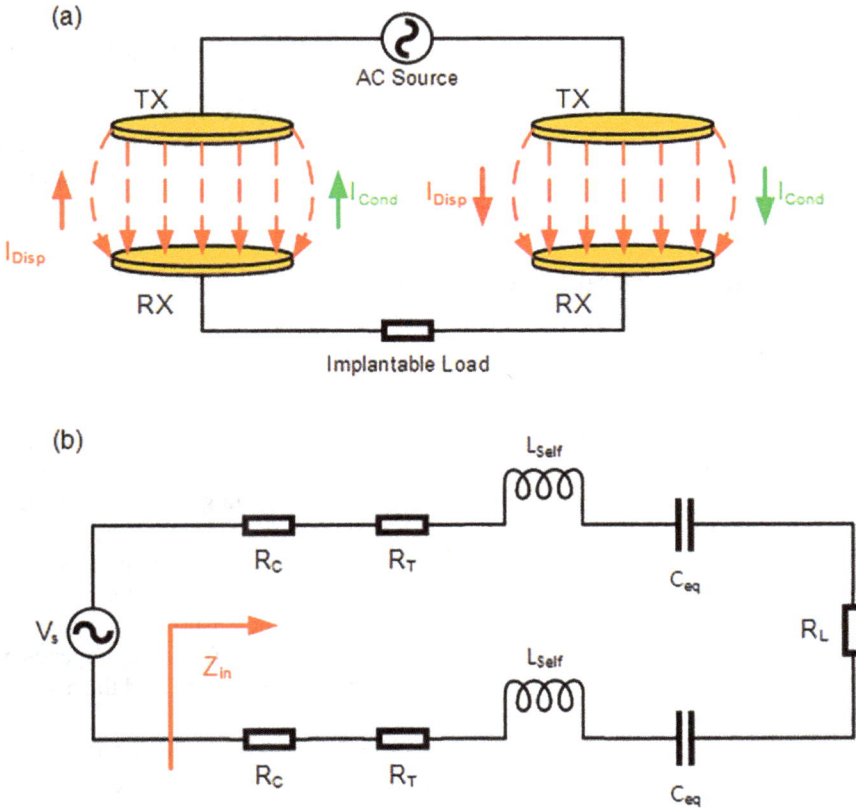

Figure 6.3 *(a) Schematic diagram and mechanism of the near-field capacitive coupling (NFCC) system [3]. (b) The NFCC link has an equivalent loss model, with tissue losses treated as series resistors (RT) and conductor losses characterized as series resistances (RC) [3]*

patches (<1 pF), resulting in power delivery compared to NRIC devices [3]. Implementing closed-loop power control, where the transmit power is changed dynamically based on real-time feedback, helps to maintain stable power delivery at the implants. Rectification becomes less efficient at frequencies higher than 30 MHz due to the limitation of RF–DC conversion. Some advanced rectification approaches, including Schottky diodes and CMOS-based rectifiers, reduce losses and increase overall efficiency. To evaluate the effect of the NFCC system on biological tissues, including SAR and the rise in temperature, a safety analysis should be performed. Electromagnetic field simulations and SAR compliance tests are necessary to ensure the biocompatibility and clinical safety of the system.

In the presence of dispersive biological tissues, the frequency range must be chosen in a way that compromises PTE and safety. Operation at high frequency reduces impedance and improves PTE at the cost of increased tissue loss. NFCC systems can be fine-tuned for safe and efficient functioning at varied implant depths and tissue conditions using optimization models and Cole–Cole parameters that quantify tissue relaxation [3].

Overall, while the NFCC system has some intriguing advantages in WPT, its sensitivity to electrode alignment, frequency tuning, and tissue loss is in dire need of strongly designed strategies and improvement within rectification techniques to overcome these inherent constraints.

6.4   Mid-field nonradiative coupling

The mid-field WPT strategy overcomes the classic constraints of both NF and FF techniques, especially for deeply implanted devices. The mid-field systems work if the TX and RX antennas are separated by about one wavelength, usually in the low-GHz frequency range. This kind of technology combines NF inductive technologies with the properties of FF radiative, hence allowing higher values of PTE for small-sized and deeply embedded implants. The schematic coupling of the mid-field is shown in Figure 6.4.

Unlike NRIC, which relies on weakly coupled coils separated by no more than a few centimeters, mid-field devices can provide higher PTE at greater distances. Careful selection of an optimal operating frequency, while ensuring safety for the tissue, can enhance power transfer performance. The operating frequency depends on the dielectric properties of human tissue, the depth of implantation, and the distance between the TX and RX parts [3,6,17].

The efficiency in power transfer for mid-field WPT systems is determined by the two-port network model, where the impedance matching between load and transmission system plays a significant role. The efficiency that can be achieved could be described as (6.5):

$$\eta \approx \frac{|Z_{21}|^2}{R_1} \frac{R_L}{|Z_{22} + Z_L|^2}. \tag{6.5}$$

The network parameters are represented by Z_{21} and Z_{22}.

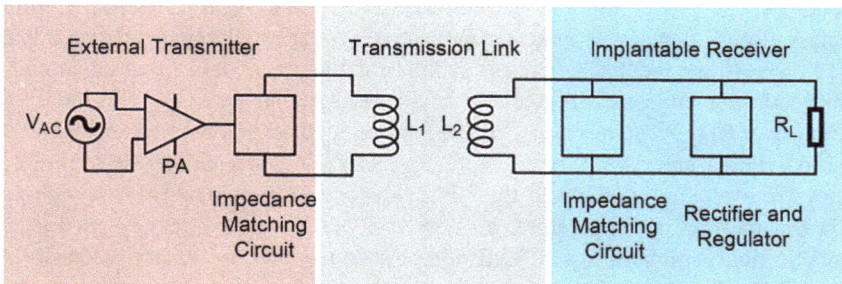

Figure 6.4   The mid-field WPT system includes the external TX portion (power amplifier and matching network), TX–RX link, and implanted RX section (matching network, rectifier, and regulator) [3]

The resistance of the load is indicated by R_L. Conjugate impedance matching, while increasing efficiency, will serve to optimize the transfer of power between the transmitter and implanted RX antenna.

The mid-field system normally operates at frequencies less than a few gigahertz. Tissue losses at these frequencies are minimal and the open-circuit voltage is low.

The voltage induced in the RX coil increases. However, dielectric losses in biological tissues impose an upper bound on the frequency for which an optimal balance between power transfer and tissue heating needs to be achieved. The maximum coupling parameter can further improve the PTE.

$$\gamma = \frac{|V_{OC}|^2}{2P_1}. \tag{6.6}$$

In this equation, V_{OC} is the open-circuit voltage and P1 is the input power. The optimal frequency selection is determined by the Debye relaxation features of human tissue, in which molecule dipole polarization contributes to energy losses.

The mid-field WPT system includes the electromagnetic interfaces for the transmitter and receiver, which are separated by a distance equivalent to the wavelength of the signal. The design is centered on two key issues:

Impedance Matching: TX and RX should have electrical impedance matching in order to reduce reflection losses and enhance power delivery. Power coupling can be maximized by using techniques like conjugate matching and current distribution optimization.

TX Source Design: The transmitting antenna should be carefully constructed to enhance electromagnetic (EM) energy coupling to the RX implant. A focused TX beam enhances focus and guarantees that transmitted waves converge efficiently at the RX point. Advanced TX beamforming techniques, such as planar immersion lenses [3,7], have been proven to improve PTE for deep embedded implants.

Although mid-field WPT offers various benefits over conventional approaches, numerous hurdles remain, notably its use to biomedical implants.

Although mid-field devices claim a higher PTE than standard NF solutions, the delivered power is often only a few microwatts (mW). Advances in ultra-low power circuits will improve energy efficiency. TX beam focusing techniques can help increase power delivery efficiency by focusing energy on the RX antenna.

Power transmission through biological tissues is always lost because of dielectric relaxation and conduction losses. Safety constraints, such as the specific absorption rate, must be maintained to avoid heating effects. These losses can be reduced by operating at an appropriate frequency and carefully modeling the tissue interactions. Electromagnetic simulations and SAR compliance studies are required to ensure safety in operations.

Very few long-term studies have been conducted on the biological effects of mid-field power transmission. More experimental validation is needed to establish the long-term safety and efficiency of mid-field systems in clinical applications.

This means that PTE, safety, and operational range must be balanced for optimal system performance. The trade-offs between operating frequency, power supply, and tissue loss need thoughtful system design. Mid-field WPT systems can achieve effective energy transfer for deep-seated implants using sophisticated antenna design, impedance matching techniques, and frequency optimization, paving the way toward more robust and reliable biomedical applications.

6.5   Far-field radiative coupling

The FF region denotes the regime of electromagnetic waves propagating as well-formed plane waves with orthogonal electric (E) and magnetic (T) fields and propagation direction. Unlike in the NF region, which is based on nonradiative methods such as magnetic inductive or electric capacitive coupling, radiative energy transmission is used in the FF. The schematic of the implantable far-field RF energy harvester is shown in Figure 6.5.

Antennas radiate energy in the form of electromagnetic waves, which is the power density that decreases with increased distance. The efficiency of the power transfer reduces by the inverse square law ($1/r^2$), where r represents the distance between transmitting and receiving antennas.

In the FF, electromagnetic waves create a stable radiative pattern in which the oscillating fields are perpendicular to each other and to the direction of propagation. Waves carry energy away from the transmitting antenna and can be absorbed by the receiving antenna to convert it into electrical power.

The FF region begins at a distance (r) from the antenna, which is roughly:

$$r > \frac{2D^2}{\lambda} \tag{6.7}$$

where D is the maximum physical size of the transmitting antenna, and λ is the wavelength of the transmitted signal. Beyond this distance, electromagnetic energy propagates largely as radiative waves; hence FF transfer is basically radiative in nature.

In a radiative energy transfer system, an RF source, either intentional or intentional, is connected to a transmitting antenna, which launches electromagnetic waves into space. A receiving antenna, positioned at some distance, captures part of the radiated power and converts it into usable energy, which is then delivered to a load,

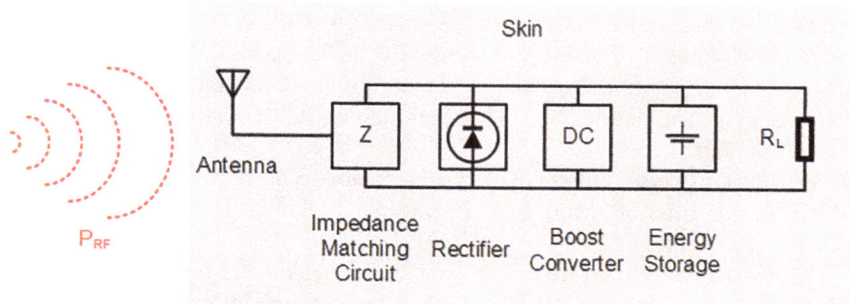

Figure 6.5   The schematic of the implantable far-field radio frequency (RF) energy harvester [5]

for example, a wireless sensor. All these are from electromagnetic radiation, and the device that receives an RF signal is properly named a rectenna-a hybrid word for "rectifier" antenna.

The power transfer between the transmitting and receiving antennas can be determined using the Friis transmission equation, which regulates the relationship between the transmitter power, the received power, the gain of the antenna and the distance. Assuming a case where the transmitting antenna radiates power isotropically, that is to say, uniformly in every direction, the power density $S(r)$ at some distance r from the transmitting antenna is given by

$$S(r) = \frac{P_T}{4\pi r^2} \tag{6.8}$$

where P_T denotes the total power provided to the transmitting antenna. This expression implies that the antenna radiates equally in all directions, resulting in a spherical wavefront. However, in reality, antennas do have directional properties, having more power radiated in certain directions while emitting less elsewhere. The antenna gain captures this directionality of the signal.

G_T increases power radiated in the desired direction. Thus, the power density $S(r)$ in the direction of maximum radiation is given by

$$S(r) = \frac{P_T G_T}{4\pi r^2} \tag{6.9}$$

The receiving antenna at distance r intercepts a portion of the radiated power with its effective aperture (A_{eR}), in square meters, which is an area over which the antenna captures energy. The P_R of the received power can be written as

$$P_R(r) = S(r)A_{eR} = \frac{P_T G_T A_{eR}}{4\pi r^2} \tag{6.10}$$

The receiving antenna does not have the same sensitivity for signals arriving from all directions; its sensitivity depends on the direction of the incoming wave. The gain of the receiving antenna G/R measures its ability to focus and capture radiation in certain directions. The effective aperture A_{eR} of the receiving antenna is related to its gain G_R by the following mathematical expression:

$$A_{eR} = \frac{G_R \lambda^2}{4\pi} \tag{6.11}$$

The wavelength of the RF signal is denoted by the symbol λ. Substituting this expression for A_{eR} into the received power equation gives the familiar Friis transmission equation:

$$P_R = P_T G_T G_R \left(\frac{\lambda}{4\pi r}\right)^2 \tag{6.12}$$

This equation describes the factors that determine the power transfer between the transmitting and receiving antennas: transmitted power P_T, antenna gains G_T and G_R, wavelength lambda, and separation distance r.

Although the Friis equation provides the basic understanding of radiative power transmission, a number of assumptions should be considered upon which the development is based. First, the equation considers that both antennas are precisely aligned in terms of gain directionality, meaning that the sending antenna's radiation is directed into the receiving antenna's most sensitive axis. In addition, the antennas have to be polarization aligned: The sent and received electric fields should be of the same orientation.

Another essential assumption is that the antennas are sufficiently separated to be in the FF of each other. In this region, the electromagnetic fields are well-constituted plane waves, and the power density decreases predictably with distance. If the antennas are too close to each other, NF processes such as inductive or capacitive coupling may dominate, and the Friis equation will not apply.

Another drawback is that G_T and G_R are not constants at all frequencies. The expressions in gains and power transfer depend on one wavelength λ, and any frequency change will alter the result. To increase the efficiency of radiative power transfer for practical systems, careful design of the antenna, alignment, and operation frequency must be performed.

As concluded, the Friis transmission equation provides a quantitative basis upon which radiative RF energy transfer can be understood. These would include optimizing factors such as antenna gain, alignment, and operational frequency in trying to increase power delivery efficiencies over long distances with massive increases, hence making the systems quite robust for WPTs.

6.6   Ultrasonic energy transfer

The principle of the ultrasonic energy transfer (UET) method involves the use of ultrasound waves (frequency >20). It is possible to implement wireless energy transmission at a frequency higher than 20 kHz. In contrast to EM technology, which may support the transmission of energy through vacuum, ultrasonic energy transfer requires a propagation medium (e.g., biological tissue). This makes UET particularly useful for IMDs, as ultrasound waves can pass through bodily tissues and give energy to deeply buried implants. The transmitted acoustic energy is captured through a piezoelectric transducer embedded inside the implant, which converts it into electrical energy. The schematic is shown in Figure 6.6.

In a typical UET system, the transmitting unit (TX) consists of an ultrasonic oscillator that is electrically energized to generate surface vibrations in the ultrasonic frequency range, typically in the range from 200 to 1.2 MHz. These oscillations generate propagating pressure waves in the medium to transport energy to the implanted receiving unit (RX). The RX is a piezoelectric energy harvester that converts acoustic energy into usable electrical power.

In view of the Huygens principle, it is possible to treat every point on the surface of a transducer as a distinct source of radiation. The final audible field can then be approximated by adding the radiation from each source. The following presents a pressure field.

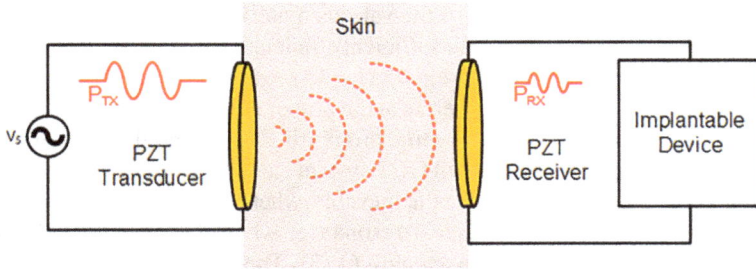

Figure 6.6   A diagram of the ultrasonic energy transfer technology. When the TX transducer is excited with voltage, it generates a pressure wave that causes a voltage to be induced across the terminals of RX. This induced voltage is rectified to provide power to the implanted device [3]

The Rayleigh integral represents $P(x, y, z; t)$ for the observation location $L(x, y, z)$ (6.13):

$$P(x, y, z; t) = \frac{jk\rho_0 c_0 u_0}{2\pi} e^{j\omega t} \int_S \frac{e^{-jkR}}{R} dS \tag{6.13}$$

In this equation, R is the distance from the source, u_0 is the amplitude of the vibration, k is the wave number, and c_0 is the wave velocity in the propagation medium.

The propagation of the ultrasonic pressure field can be divided into three main regions: Near Field (NF): The oscillations occur in this field with several maxima and minima, which makes the power supply not certain. Far Field (FF): Beyond the natural focus zone, the acoustic pressure decays twice as fast as the distance, so this region is not advantageous for energy transfer. Natural Focus Zone: The pressure field converges to a natural focal point, thus allowing maximal energy delivery to the RX. Distance L from the natural focus (6.14):

$$L = \frac{(2a)^2 - X^2}{4\lambda} \approx \frac{a^2}{\lambda} \tag{6.14}$$

where α is the radius of the transducer and X is the observation distance. The natural focus is the best spot for implanting the RX, allowing the TX to transfer the most acoustic power possible.

The PTE of ultrasonic energy transfer devices is controlled by tissue attenuation, acoustic impedance mismatches, and the mechanical-to-electrical conversion efficiency. The effectiveness of a piezoelectric transducer receiver is mainly determined by the quality factor Q and the coupling coefficient of the material k. For example, lead zirconate titanate (PZT) transducers are more efficient than polymer-based equivalents such as PVDF [7].

Impedance mismatches at the tissue–transducer interface will cause acoustic reflections that limit the overall amount of power delivered to the RX. The pressure reflection coefficient (τ) at the interface can be expressed as (6.15) [7]:

$$|\tau| = \left| \frac{Z_{\text{tissue}} - Z_{\text{PZT}}}{Z_{\text{tissue}} + Z_{\text{PZT}}} \right| \tag{6.15}$$

where Z is the acoustic impedance of the various materials.

Power conditioning circuits with sufficient efficiency are a prerequisite to achieve maximum overall performance. The circuit should be designed in such a way that the piezoelectric transducer is able to operate at the proper resonance frequency at the RX for an efficient drawing of power out while prohibiting the harmonic mode excitation.

In general, designing a successful UET system calls for several crucial elements: Maximum acoustic power transfer is achievable when both transducers operate at their resonance frequency. Mechanical resonance, when the thickness of a transducer is an odd multiple of half-wavelength ($\lambda/2$); There should be proper acoustic impedance matching between transducers and tissues to reduce reflections and enhance power delivery. The frequency range of 200 kHz to 1.2 MHz has to be selected based on the balance between energy loss due to tissue attenuation and efficiency. With increased frequency, the transducer size decreases, while tissue absorption increases, which limits the range of transmission. Variations in tissue characteristics (such as moisture and temperature) and relative motion between the TX and the RX can induce power fluctuations. Closed-loop power regulation solutions can help prevent these effects while also stabilizing system performance [3].

There are a number of challenges that must be overcome for ultrasonic energy transfer to take place in implantable devices. Soft tissues absorb ultrasonic waves between 0.6 and 1.5 dB/cm (at 1 MHz), so their effective range is minimized. Possible ways to improve that include optimizing operating frequency and finding implant positions with minimal attenuation. Continuous ultrasonic exposure can cause tissue vibration or heating, from which the FDA safety limits should not be exceeded, for example, an intensity limit of 7.2 W/mm^2. Body movement and alteration of tissue characteristics can cause system alignment and variation in efficiency. Variations that can also be compensated for in real time using closed-loop feedback devices. UET systems produce consistent power delivery, but most of their overall power output has been restricted to a couple of mW, hence UET systems have traditionally been used with low-power implants. Advances in ultra-low power circuits will help mitigate this barrier to system practicality [3].

Ultrasonic energy transmission has huge potential to power deeply embedded implants, including microoxygen generators, electrostimulators, and neurological implants. The ability to transfer power through biological tissues, together with developments in transducer design and power conditioning, makes the technique a very attractive option besides conventional EM-based treatments. Further research on safety and long-term effects and system miniaturization will open up wider clinical applications.

6.7   Harvesting vs. wireless power transfer

For the harvesting of RF energy to be practical, the power demands of the proposed application must be compared with the amount of RF energy that can be practically harvested from the surrounding environment. Commercially available wireless sensor nodes have typical power requirements of around 100 µW and reception antennas with feasible dimensions of only a few square centimeters [5].

Several devices operating in widely available frequency bands have been identified as promising candidates for RF energy harvesting in urban environments. These include GSM900 (downlink: 935–960 MHz), GSM1800 (downlink: 1805–1879 MHz) and internationally available Wi-Fi frequencies (2.4–2.5 GHz). These devices are common for RF signal harvesting and can accept antenna sizes of 10 to 50 cm^2 [25].

Power density measurements have given some insight into the energy that is available within these frequency bands. GSM 900 downlink frequency surveys indicate that the power density ranges from 0.01 to 0.3 μW/cm^2 depending on the distance from the base station (e.g. 25–100 meters). Power levels similar to these have also been measured in the GSM 1800 band; however, Wi-Fi signals have power densities nearly an order of magnitude lower, which further limits accessible energy for harvesting [18].

The follow-up points to a critical limit related to ambient RF energy harvesting-the levels of power that could feasibly be available remain very small, incomparable even to the modest 100 μW that the majority of today's wireless nodes require. Only amazingly huge reception antennas in sizes such as 330–1000 cm^2 would produce that kind of power, which again is unreal for most applications in reality.

Because of this limitation, only ambient RF energy-based systems cannot be used in applications requiring periodic and reliable power output; instead, the use of RF energy transfer technologies is necessary. In such a technology, an RF source specifically designed for that purpose intentionally transmits power to the receiving antenna to ensure sufficient energy delivery.

In practice, RF energy transfer systems use license-free ISM frequency bands such as 868 MHz in Europe, 900 MHz in North America, and 2.4 GHz globally. These frequencies are controlled such that the maximum permissible transmission power, which is commonly referred to as the Effective Isotropic Radiated Power (EIRP), is limited. EIRP represents the total power that an isotropic radiator would radiate uniformly in all directions. It is calculated by multiplying the transmitter power P_t by the antenna gain G_t.

$$P_t G_t = EIRP \tag{6.16}$$

In some circumstantial circumstances, this (6.16) is further related to effective radiation power (ERP), particularly for half-wave dipole antennas, where

$$EIRP = 1.64 ERP \tag{6.17}$$

The concept of RF energy harvesting from ambient sources is very attractive; however, the low power density of accessible signals, especially in urban environments, makes it unsuitable for applications requiring continuous power at the μW level without very large antennas. In contrast, a more reliable method for wireless power supply to small devices is represented by specialized RF energy transfer systems operating inside controlled ISM frequency bands. With proper adaptation of transmission power, antenna design, and operating frequency, these systems may mitigate the limitation of ambient RF energy harvesting and ensure a consistent transmitted energy.

References

[1] S.-E. Adami, P. Proynov, G.S. Hilton, *et al.* A flexible 2.45-ghz power harvesting wristband with net system output from 24.3 dbm of RF power. *IEEE Transactions on Microwave Theory and Techniques*, 66(1):380–395, 2017.

[2] A. Afsahi, B. Afshar, E. Afshari, *et al.* 2010 index IEEE Journal of Solid-State Circuits, Vol. 45. *IEEE Journal of Solid-State Circuits*, 45(12):2883, 2010.

[3] K. Agarwal, R. Jegadeesan, Y0-X. Guo, and N.V. Thakor. Wireless power transfer strategies for implantable bioelectronics. *IEEE Reviews in Biomedical Engineering*, 10:136–161, 2017.

[4] D.A. Borton, M. Yin, J. Aceros, and A. Nurmikko. An implantable wireless neural interface for recording cortical circuit dynamics in moving primates. *Journal of Neural Engineering*, 10(2):026010, 2013.

[5] D. Briand, E. Yeatman, and S. Roundy, editors. *Micro Energy Harvesting*. Wiley-VCH, 2015.

[6] S. Gabriel, R.W. Lau, and C. Gabriel. The dielectric properties of biological tissues: III. Parametric models for the dielectric spectrum of tissues. *Physics in Medicine & Biology*, 41(11):2271, 1996.

[7] J.S. Ho, B. Qiu, Y. Tanabe, A.J. Yeh, S. Fan, and A.S.Y. Poon. Planar immersion lens with metasurfaces. *Physical Review B*, 91(12):125145, 2015.

[8] I. Hochmair, P. Nopp, C. Jolly, *et al.* MED-EL cochlear implants: State of the art and a glimpse into the future. *Trends in Amplification*, 10(4):201–219, 2006.

[9] W.G. Holcomb, W.W.L. Glenn, and G. Sato. A demand radiofrequency cardiac pacemaker. *Medical and Biological Engineering*, 7(5):493–499, 1969.

[10] R. Jegadeesan, K. Agarwal, Y.-X. Guo, S.-C. Yen, and N.V. Thakor. Wireless power delivery to flexible subcutaneous implants using capacitive coupling. *IEEE Transactions on Microwave Theory and Techniques*, 65(1):280–292, 2016.

[11] R.-H. Kim, H. Tao, T.-i. Kim, *et al.* Materials and designs for wirelessly powered implantable light-emitting systems. *Small*, 8(18):2812–2818, 2012.

[12] T.-i. Kim, J.G. McCall, Y. Hwan *et al.* Injectable, cellular-scale optoelectronics with applications for wireless optogenetics. *Science*, 340(6129):211–216, 2013.

[13] A. Kurs, A. Karalis, R. Moffatt, J.D. Joannopoulos, P. Fisher, and M. Soljacic. Wireless power transfer via strongly coupled magnetic resonances. *Science*, 317(5834):83–86, 2007.

[14] C. Liu, Y. Zhang, and X. Liu. Circularly polarized implantable antenna for 915 MHz ISM-band far-field wireless power transmission. *IEEE Antennas and Wireless Propagation Letters*, 17(3):373–376, 2018.

[15] I. Mayordomo, T. Dräger, P. Spies, J. Bernhard, and A. Pflaum. An overview of technical challenges and advances of inductive wireless power transmission. *Proceedings of the IEEE*, 101(6):1302–1311, 2013.

[16] M.H. Ouda, M. Arsalan, L. Marnat, A. Shamim, and K.N. Salama. 5.2-GHz RF power harvester in 0.18-μm CMOS for implantable intraocular

pressure monitoring. *IEEE Transactions on Microwave Theory and Techniques*, 61(5):2177–2184, 2013.

[17] A.S.Y. Poon, S. O'Driscoll, and T.H. Meng. Optimal frequency for wireless power transmission into dispersive tissue. *IEEE Transactions on Antennas and Propagation*, 58(5):1739–1750, 2010.

[18] A. Shameli, A. Safarian, A. Rofougaran, M. Rofougaran, and F. De Flaviis. Power harvester design for passive UHF RFID tag using a voltage boosting technique. *IEEE Transactions on Microwave Theory and Techniques*, 55(6):1089–1097, 2007.

[19] Y.-C. Shih, T. Shen, and B.P. Otis. A 2.3 μW wireless intraocular pressure/temperature monitor. *IEEE Journal of Solid-State Circuits*, 46(11):2592–2601, 2011.

[20] A.M. Sodagar and P. Amiri. Capacitive coupling for power and data telemetry to implantable biomedical microsystems. In *2009 4th International IEEE/EMBS Conference on Neural Engineering*, pages 411–414. IEEE, 2009.

[21] D. Tan. IEEE Journal of Emerging and Selected Topics in Power Electronics receives high ranking in impact factor [Society News]. *IEEE Power Electronics Magazine*, 4(3):80–80, 2017.

[22] N. Tesla. Apparatus for transmitting electrical energy, December 1, 1914. US Patent 1,119,732.

[23] L.-G. Tran, H.-K. Cha, and W.-T. Park. RF power harvesting: A review on designing methodologies and applications. *Micro and Nano Systems Letters*, 5(1):14, 2017.

[24] L.-G. Tran, H.-K. Cha, and W.-T. Park. RF power harvesting: A review on designing methodologies and applications. *Micro and Nano Systems Letters*, 5(1):1–16, 2017.

[25] H.J. Visser and R.J.M. Vullers. RF energy harvesting and transport for wireless sensor network applications: Principles and requirements. *Proceedings of the IEEE*, 101(6):1410–1423, 2013.

[26] C.T. Wentz, J.G. Bernstein, P. Monahan, A. Guerra, A. Rodriguez, and E.S. Boyden. A wirelessly powered and controlled device for optical neural control of freely-behaving animals. *Journal of Neural Engineering*, 8(4):046021, 2011.

[27] R.-F. Xue, K.-W. Cheng, and M. Je. High-efficiency wireless power transfer for biomedical implants by optimal resonant load transformation. *IEEE Transactions on Circuits and Systems I: Regular Papers*, 60(4):867–874, 2012.

[28] M. Zargham and P.G. Gulak. Maximum achievable efficiency in near-field coupled power-transfer systems. *IEEE Transactions on Biomedical Circuits and Systems*, 6(3):228–245, 2012.

[29] M. Zimmerling. MRI-safe implant magnet with angular magnetization, January 23, 2018. US Patent 9,872,993.

Chapter 7
Photovoltaic energy harvesting in biomedical implantable devices

Wireless implantable technologies are becoming more common in biomedical applications such as physical identification, real-time health monitoring, and physiological trait recording. Current implantable devices, which frequently require surgical replacement, are powered by batteries. Self-powered implanted devices are appealing in this sense for real-time monitoring of human physiological features. Furthermore, electricity collection and generation beneath human tissue remains a serious hurdle. Nonetheless, as packaging and nanotechnology improve, alternate harvesting approaches based on piezoelectricity, thermoelectricity, biofuel, and radio frequency power transfer are emerging. All of these approaches have constraints such as limited power output, bulky size, or low efficiency. Photovoltaic (PV) energy conversion is one of the most promising candidates for implantable applications due to their higher-power conversion efficiencies and small size. Because of its greater conversion efficiencies and small size, PV energy conversion is one of the most attractive possibilities for implantable applications.

7.1 Background of photovoltaic effect

PV effect is the process to directly convert the light in to electricity. Nowadays, it is an increasing renewable alternative resources according to the conventional fuel consumption [6]. The first PV cell was manufactured in 1950s, and the technology was tremendously developed around 1960s because of space explore. Actually the high fabrication cost of PV cell makes it hard in electricity generation in industry. However, after experiencing fuel shortage in the 1970s, the industry started concentrating on the terrestrial application of PV cells rather than their use in the space to overcome the following energy shortage, which promoted the investigation of PV cell as a new option to generating terrestrial power. However, it is still a challenge to apply PV cell in grid power generation [6]. PV cell is an electronic device which is based on PV effect to convert light into electricity. Light illuminating on the PV cells produces a voltage and a current to generate electricity. To achieve the electricity, first a special material is required to absorb photons and to raise the electrons from lower energy state to higher energy state. The electrons in higher energy state move from the PV cell to an external circuit or load. The energy from these electrons will be dissipated in the

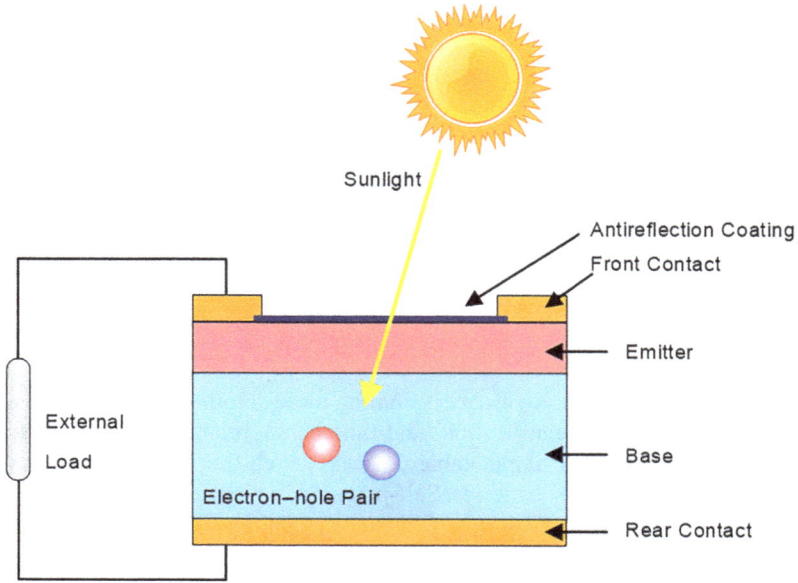

Figure 7.1   p–n Junction PV cells and the theory of electricity generation from PV effect

external circuit, and the electrons will return PV cell with a low energy state [6]. The PV cell structure is shown in Figure 7.1. There are three generations of PV cell in the industry up to date. The first generation of PV cell is manufactured using monocrystalline and polycrystalline silicon, which together account for 90% of the PV cell market due to the high power conversion efficiency [92]. The second generation PV cell is based on the thin-film technology in silicon. The material of amorphous silicon and microcrystalline silicon are invented, which improved the flexibility and decreased the fabrication cost of the PV cells by using higher extinction coefficient and thinner material. The latest and third generation of PV cells is using emerging polymer and organic materials, which can make the cost much lower than the crystalline or thin-film-based PV cells. However, they still need more investigation to improve their power conversion efficiency and stability [92].

7.1.1   The theory of light radiation

For implantable power harvesting applications, PV cells cannot harvest in vivo bioenergy directly, but it can harvest energy from an ambient light source (natural light or artificial light) [35]. The amount of energy that the Sun emits and incident on the Earth is much greater than what we see on a daily basis. The visible light that humans see from the Sun represents a small portion of the electromagnetic (EM) spectrum, which is depicted to the right. The EM spectrum illustrates light as a wave related to a specific wavelength. The EM spectrum of the solar radiation is shown in

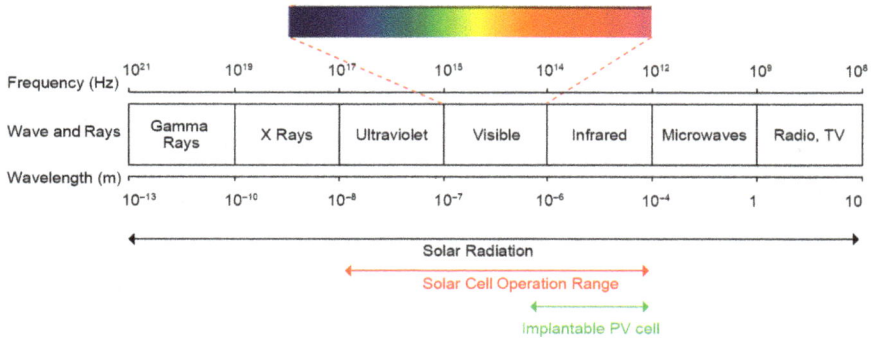

Figure 7.2   The electromagnetic spectrum of solar radiation, and the operation range of normal cell and the implantable PV cell

Figure 7.2 [78,98]. The interaction of incident sunlight with a PV converter or any other item depends on a number of vital aspects of the incident solar energy. The vital characteristics of the incident solar energy are

- the spectrum of the input light,
- solar radiation power density,
- the angle where incident solar light impacts the PV cells, and
- the solar radiation received by a specific surface over a specific time.

For an implantable PV cell, the transmittance of the skin layer on top of the PV cell should be also considered.

In addition to spectrum irradiance, photon flux is frequently required for the examination of solar cells. In order to obtain the spectral irradiance from the photon flux, convert the photon flux at a specific wavelength to W/m^2. Atmospheric factors including absorption, reflection, and scattering have numerous impacts on the solar radiation at the Earth's surface. The length of the light's travel through the atmosphere, normalized to its shortest path length, is the air mass (AM). The AM determines the reduction of the radiation from solar light. The majority of PV cell is measured and operated under AM 1.5G. Despite the natural light, artificial light is also applied in powering implantable PV cells to obtain optimal optical transmittance from the skin.

7.1.2   The mechanism of the incident light

When the incident light propagates in the materials, part of the light is absorbed, part of the light is reflected, and part of the light is transmitted (refracted and scattered), which is shown in Figure 7.3(a). Besides, Snell's law and Maxwell's equation to analyze light penetration can be indicated by Jones' Matrix [59].

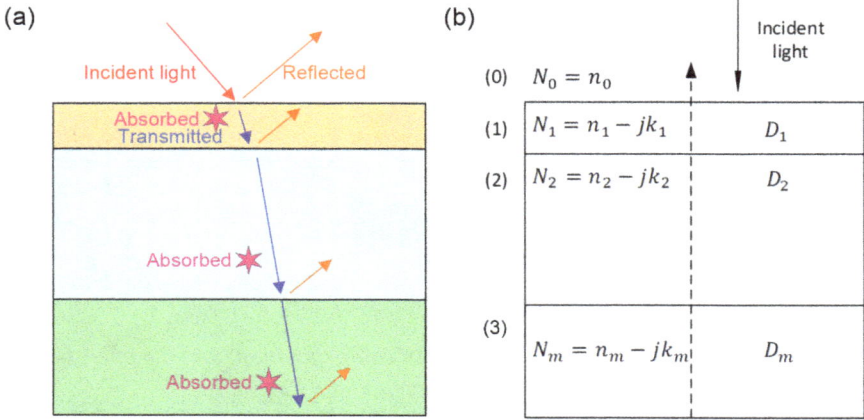

Figure 7.3   Panel (a) shows the light absorption, reflection, and transmission in different materials. Panel (b) shows the schematic of Jones Matrix.

To evaluate the optical characteristics in 2D structure, the 2×2 Jones Matrix was applied [9], which is also shown in Figure 7.3:

$$\begin{bmatrix} E_{Ra} \\ B_{Ra} \end{bmatrix} = \sum_{i=1}^{m} \begin{bmatrix} \cos \phi_i & \left(\frac{i}{\eta_i}\right) \sin \phi_i \\ j\eta_i \sin \phi_i & \cos \phi_t \end{bmatrix} \begin{bmatrix} 1 \\ \eta_m \end{bmatrix} \tag{7.1}$$

$$M_{total} = M_{0,1} \times M_{1,2} \times M_{2,3} \times \dots \times M_{m-1,m} \tag{7.2}$$

where δ_i is the wave phase shift ($\delta_i = 2\pi N_i d_i \cos\theta i / \lambda$) in the ith layer, N_i is the refractive index, d_i is the thickness of the ith layer, and η_i is the pseudo index in i^{th} layer ($\eta_i = N_i \cos\theta_i$). E_{Ra} or B_{Ra} is the ratio between the electric and magnetic fields of the transmitted light and incident light. M is the total number of layers. The reflectance (R), absorptance (A), and transmittance of light can be evaluated by calculating the components in M_{total} [9]:

$$R = \frac{4\eta_0 Re\left(\eta_m\right)}{\left(\eta_0 E_{Ra} + B_{B_n}\right)\left(\eta_0 E_{Ra} + B_{Ra}\right)^*} \tag{7.3}$$

$$A = \frac{4\eta_0 Re\left(E_{Ra}B_{Ra}^* - \eta_m\right)}{\left(\eta_0 E_{Ra} + B_{B_0}\right)\left(\eta_0 E_{Ra} + B_{Ra}\right)^*} \tag{7.4}$$

$$T + A + R = 1 \tag{7.5}$$

7.1.3   The theory of semiconductor

As mentioned previously, the PV cell is based on the PV effect. Most PV cells are fabricated with semiconductor materials such as Si, GaAs, and CdTe. Organic materials are generating much interest due to their flexibility, low price, and low fabrication costs, while they usually have a large active area and low stability [103]. Most implantable PV cells are made of crystalline silicon due to their high efficiency and nontoxicity [103]. The composition of the semiconductor is analyzed by invoking Poisson equation and continuity equation. Besides, I invoke Snell's law and Maxwell equation to analyze light penetration, which can be indicated by Jones' matrix [59]. Poisson's equation indicates the electric potential produced by given charges and mass density distribution [5]:

$$-\nabla \cdot \nabla \psi = \frac{q}{\varepsilon} \left(p - -n + N_D^+ - N_A^- \right) \tag{7.6}$$

where n and p are the electron and hole concentrations, ψ is the electrostatic potential, q is the electron charge, ε is the permittivity of semiconductor, and N_D^+ and N_A^- are the ionized donor and acceptor impurity concentrations, respectively. To analyze the carrier transport and the gradients of the parameters in material, I enable the Continuity Equations with drift and diffusion as follows [5]:

$$\frac{\partial n}{\partial t} = -\mu_n n \nabla^2 \psi + D_n \nabla^2 n + G - R$$
$$\frac{\partial p}{\partial t} = \mu_p p \nabla^2 \psi + D_p \nabla^2 p + G - R \tag{7.7}$$

where k is Boltzmann constant, T is the temperature of ambient, μ_n and μ_p are the electron and hole mobilities, respectively, $D_{n,p}$ are the electron and hole diffusion coefficient ($D_{n,p} = \mu_{n,p} kT/q$), and G and R present the generation rate and recombination rate of electron–hole pairs, respectively. Direct radiative recombination (7.8) depends on dopant (p, n), injection factor (g_{eh}), and radiative coefficient (B). Auger recombination (7.8) dominates at high dopant densities, and it also depends on the Auger coefficient (C_n and C_p), injection factor (g_{eeh} and g_{ehh}), and dopant. Shockley–Read–Hall (SRH) recombination via states in the forbidden band is shown in (7.8), which is influenced by recombination lifetime ($\tau_{p,n}$). The recombination rates are as follows [5]:

$$R_{SRH} = \frac{pn - n_i^2}{\tau_p(n+n_1) + \tau_n(p+p_1)}$$
$$R_{Au} = \left(g_{eeh} C_n n + g_{ehl} C_p p \right) \left(pn - n_i^2 \right) \tag{7.8}$$
$$R_{rad} = g_{eh} B \left(pn - n_i^2 \right)$$

The generation rate indicates the generation of electron–hole pairs correlated to the absorption of photons [59]. It is related to the absorption coefficient of silicon $\alpha(\lambda)$, photon flux on the surface $b(\lambda)$, power flow according to wavelength, and the depth into the device in x, y, z-direction. The photon flux can be calculated from light intensity of AM 1.5 Global ($AM1.5G$). The generation rate can be determined by [59]:

$$G(\lambda) = \alpha(\lambda) b(\lambda) P_s(x, y, z, \lambda) \tag{7.9}$$

The electron–hole pair generation and recombination in the semiconductor are shown in Figure 7.4. The absorption coefficient is not constant, but it highly depends on wavelength of the incident light. The possibility of absorbing a photon is based on the probability of interaction between a photon and an electron to move from one energy band to another. For the photons that are with an energy close to the band gap, the absorption is relatively low because only the direct electrons at the edge of

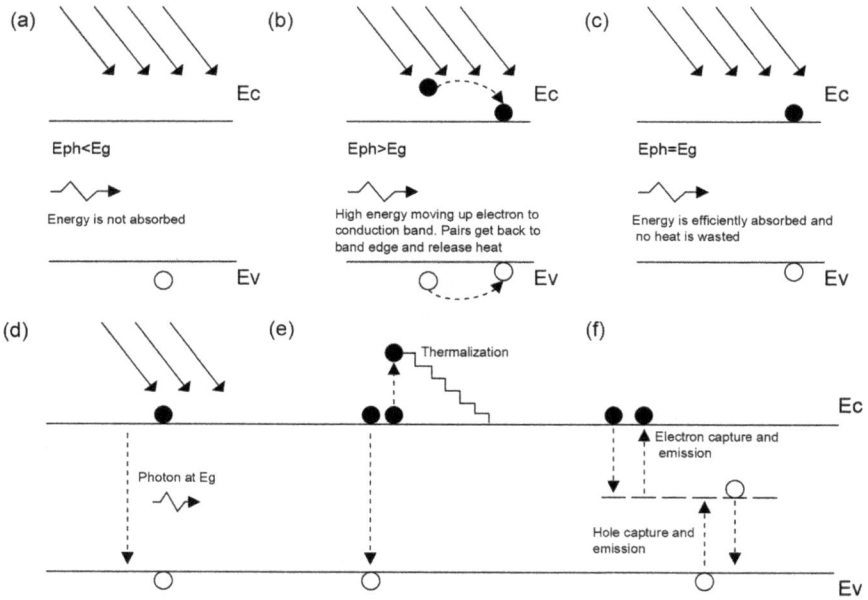

Figure 7.4   *Panels (a), (b), and (c) show the three different conditions of electron–hole pair generation, where (a) $E_{ph} > E_g$, (b) $E_{ph} > E_g$ and (c) $E_{ph} > E_g$. Panels (d), (e), and (f) show the three types of recombination in the semiconductor, where (d) is radiative recombination, (e) auger recombination, and (f) SRH recombination.*

valence band are able to react with the photon to make absorption occur. With the incident photon energy rising, not just the electrons already with energy close to the band gap can react with the incident photons. In this case, much more electrons can interact with the photon and inevitably lead to tremendously more absorption of the photon.

The absorption coefficient $\alpha(\lambda)$ is related to the extinction coefficient (k) and wavelength of the incident light; the equation is shown as follows [1]:

$$\alpha(\lambda) = \frac{2\pi k}{\lambda} \tag{7.10}$$

The absorption depth is inverse of the absorption coefficient, which illustrates how deep incident light can penetrate into a material before the light is being absorbed. The light with shorter wavelength containing higher energy and shorter absorption depth than the light with a longer wavelength. Absorption depth highly aspects of solar cell design including the thickness of the device and selection of the semiconductor material. The absorption depth of silicon according to wavelength is shown in Figure 7.5

The light-generated current is proportional to the generation rate and according to the optical properties of the material. Compared with the current, the voltage is more

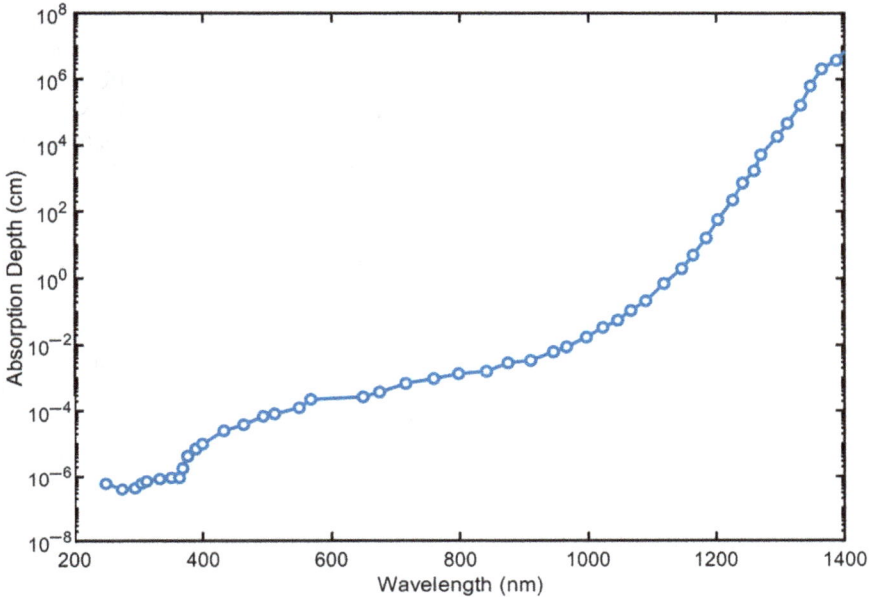

Figure 7.5   The absorption depth of the crystalline silicon according to different wavelength of the incident light [96]

related to the material properties such as doping concentration and diffusion length that slightly change with incident light. In the PV industry, the current means the ability to trap and absorb light in the device, which contributes to the high efficiency. However the maximum output is limited by the parameter called filling factor which is defined by material. Normally, maximum voltage is more influence by filling factor compared with maximum output current.

It is promising to involve the PV cell into implantable application. First, the PV cell fabrication is in a mature industry compared with the other implantable energy harvesting technology. Second, there are abundant light source for PV cell to harvest in human society, including solar light as well as artificial light. Third, the energy conversion efficiency of PV cell is much higher than the other energy harvesting technologies. Lastly, the PV cell is still under development with emerging material and technologies, which will improve the device miniaturization, flexibility, and energy conversion efficiency and obtain increasingly lower fabrication cost.

7.2   Classification of photovoltaic cell

Harvesting energy from light has been used for powering portable consumer products. Here, PV cells are used to convert light or the Sun's energy into useful electricity. Ultimately, semiconductor materials are commonly used for the purpose of producing currents and voltages as a result of the absorption of light, which is a phenomenon known as the PV effect. The PV effect is discovered in 1839 by Alexandre-Edmund Becquerel, when he observed that electric current is generated from a light-induced chemical reaction. Following that, Adam and Day discovered the PV effect in selenium in 1876 [25]. The theory of PV effect is developed by Plank and Wilson, who donate the quantum nature of light and quantum nature of the solid structure [25]. However, this phenomenon was not noticed by the industry until the first PV cell with an efficiency of 6% was invented in 1954 [25].

The first generation of PV cell is made from crystalline semiconductors, which consist of monocrystalline and polycrystalline silicon. In its most basic form, a PV cell consists of a p–n junction diode. Typical PV cell efficiencies range from 18% for polycrystalline to 24% from highly efficient monocrystalline technologies [32]. This technology constantly developed with improved capability and efficiency over a few decades. The monocrystalline structure occupies 80% of PV cell market, which essentially apply p–n junctions (shown in Figure 7.6(a)). Nowadays, the efficiency of the crystalline semiconductor is limited by the energy induced by photons because it decreases in higher wavelength [25]. In addition, the radiation with higher wavelength will heat up the temperature of the crystalline PV cells hence causes power

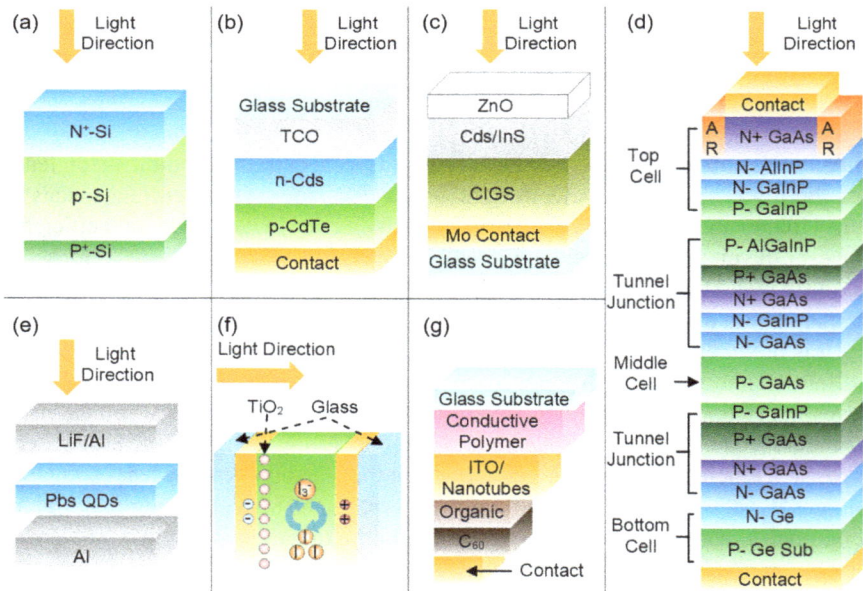

Figure 7.6   *The material and structure of the PV cell. Panel (a) shows the PIN structure of crystalline Si PV cell. Panel (b) shows the structure of CdTe thin-film PV cell. Panel (c) shows the structure of CIGS thin-film PV cell. Panels (d) shows the structure of triple-junction PV cells. Panel (e) shows the structure of quantum-dot PV cells. Panel (f) shows the structure of dye-sensitized solar cell. Panel (g) shows the structure of organic PV cells.*

dissipation [25]. The efficiency of the crystalline silicon PV cell is up to 26.7%, and the V_{oc} can be as 0.74 V measured by National Institute of Advanced Industrial Science and Technology (AIST) [37]. The other material such as GaAs has the efficiency of 28.8% and the V_{oc} of 1.12 V. as measured by National Renewable Energy Laboratory (NREL). The increased production and reduced costs in the PV industry led to the new crystallization technologies. The polycrystalline silicon was highly developed when the price of silicon was over 300$/kg. However, the efficiency of the polycrystalline Si is lower than that of monocrystalline Si, which is 15%. The other advantage of polycrystalline structure rather than monocrystalline structure is decreasing the flaws at crystal structure and metal contamination [63].

Compared with the crystalline technology, thin-film technology offers advantages such as lower costs and less environment hazards during fabrication. Unlike the structure of single crystalline semiconductor, the thin-film PV cells are normally deposited in a substrate of glass, which can decrease the deposited layer within 10 μm. In this case, the costs of the material and manufacture costs according to throughout deposition process are lowered. The thin-film PV cells contain amorphous silicon (α-Si), heterojunction structure of cadmium telluride (CdTe), and copper indium gallium selenide (CIGS) Compared with crystalline silicon, the α-Si is with higher bandgap of 1.7 eV than 1.1 eV, which leads to higher absorption in visible spectrum. The schematic of the CdTe is shown in Figure 7.6(b). The bandgap of CdTe is around 1.45 eV, which leads to an efficiency of 15% [37,38]. The development of CdTe PV cell is limitied regarding the toxic nature of the material and polution of manufacture. In addition, the material constraints of Te increase the costs of CdTe PV cell. CIGS contains the elements from group I, III, and IV in periodic table, which attributes to high absorption coefficients and electrical characteristics. Unlike the p–n junction structure, CIGS PV cells are known as heterojunction structure, which is normally deposited by sputtering "ink" printing and electroplating [26]. The schematic of CIGS PV cell is shown in Figure 7.6(c).

The multijunction PV cell is shown in Figure 7.6(d), which stacks different crystalline layers in order to absorb most of the radiation [63]. As each crystalline layer has different bandgap and absorption coefficient, the layers in the PV cell can be tuned more efficient and suitable for a wider spectrum. The triple junction PV cell can achieve an efficiency of 39% measured by AIST [37]. Originally, the whole crystalline layers are deposited in GaAs substrate because it is matched with GaAs lattice. However, germanium (Ge) substrate is mostly applied to replace GaAs substrate and reduce the cost. The first cell works as a silicon p–n junction cell. Then the GaInP and GaAs layers are deposited on the top to enhance the bandgap and capture optical energy in wider spectrum, thereby improving the efficiency.

Quantum-dot PV cells are very efficient with their distinctive absorption and emission spectra, which vary according to the size of the particles [63]. The quantum dots are in nanometer scale and have an adjustable bandgap. The strong coupling of quantum dots is required to enhance the lifetime of excitation and to provide electricity in high voltage. The schematic of quantum-dot PV cells is shown in Figure 7.6(e).

The structure of a dye-sensitized PV cell is shown in Figure 7.6(f). Such structure consists of a semiconductor layer, an electrolytic liquid produced by dissolving a salt in a solvent liquid [63]. The semiconductor and the electrolytic liquid work together to split the electron–hole pairs when the photons are absorbed. A nanomaterial such as titanium dioxide (TiO_2) is commonly applied to hold the dye molecules. As the solvent liquid is toxic and hazardous to environment, the solvent-free dye-sensitized PV cell is developed but its efficiency is low.

The structure of organic PV cells is shown in Figure 7.6(g). The organic PV cell is composed by thin-film organic semiconductor within hundreds of nanometers, such as polymer and small-molecule compounds. Typically, the organic cell contains a glass substrate, a transparent contact of indium tin oxide (ITO), a conducting polymer layer, a photoactive polymer layer, and a back metal contact layer, such as Al. As ITO is expensive in the market, the carbon nanotubes are normally used alternatively [83]. The efficiency of organic PV cells is less than 10% but cost of material and fabrication is very promising [37].

7.3   Development of implantable photovoltaic cell

In this section, I initially demonstrate the development of implantable PV cells. I also compare PV cell performance using different materials and technologies. Next, the fabrication process of PV cells and the encapsulation techniques will be discussed. Here, it is important to emphasize that encapsulation involves protecting the PV cells as well as reducing the adverse side effects to humans. I mainly discuss the transmissivity, flexibility, and biocompatibility of the encapsulation material. Optical losses due to tissue-scattering effects will be analyzed and the optical properties of different skins will also be summarized.

Concerning the development of implantable PV cells, several improvements have been made with respect to materials, device configuration, encapsulation, and stability. The major milestones of these developments are shown in Figure 7.7. Before 2012, implantable PV cells occupied a larger area due to the complementary metal-oxide-semiconductor (CMOS) technology available at the time. Between 2010 and 2014, implantable PV cell technology developed according to the involvement of the CMOS technology [11]. The on-chip PV cell was invented during this period [11,56]. Between 2014 and 2018, the literature showed a shift toward lower power density PV cells. Perhaps this could be attributed to an increased commitment toward better biocompatibility and encapsulation issues. After 2018, emerging technologies such as organic PV cells were demonstrated in implantable applications [87] without encapsulation.

Among the first researchers to propose the use of PV cells for implantable applications was Dan Tchin-iou in 1999 [22]. A source providing light in the visible range was used to power a commercial solar cell with an integrated battery pack. This system was used to power an artificial heart. However, this research lacked analysis of tissue loss and under-skin device validation (in vitro or in vivo). In addition, the light source was confined to wavelengths in the visible range and did not consider light in the red and infrared ranges.

The first in vivo testing of an implantable PV cell was achieved by Thomas Laube in 2004 [55]. A near-infrared (NIR) light source was used and PV arrays were encapsulated with a resin. The PV cells were embedded in an intraocular microsystem,

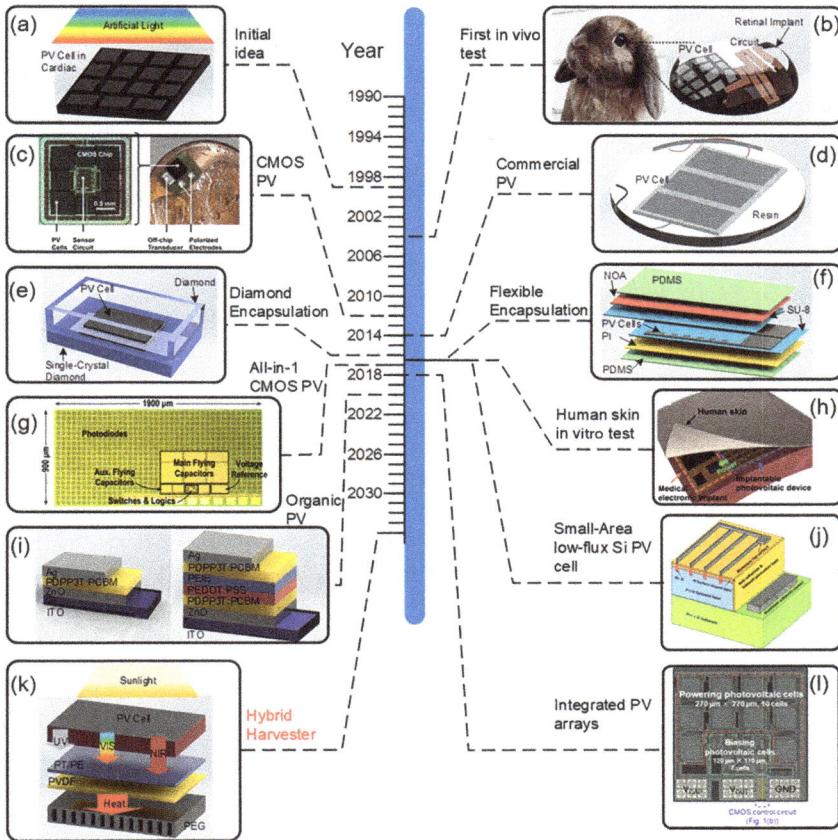

Figure 7.7   *The evolution of implantable PV cells between 1999 and 2030. The implanted PV cells have undergone 20 years of development to get from a proposed ideal to a refined approach that may be tested in vivo or in vitro. Also displayed is a 10-year development prospection (2020–30). The initial design for an implanted PV cell is shown in (a). The implantable PV cell's initial in vivo test on an animal is shown in figure (b). (c) The first CMOS PV cells on-chip. (d) Powering pacemakers with commercial PV cells. (e) The implanted PV cell is enclosed in a diamond. (f) Multilayer flexible encapsulations are used to create flexible PV cells. (g) CMOS PV cells that are stacked and have a power management system. (h) The first human skin implanted PV cell test. The original organic implantable PV cells (single-junction and tandem Cells). (j) Low-flux implantable PV cell with a small surface area. (k) Prognostication of hybrid PV with additional energy harvesters. (l) Bonding of integrated silicon PV panels to CMOS circuit.*

and the whole chip was implanted into a rabbit. The system was tested in vivo for 7 months, which is the longest testing duration for any implantable device to date. In addition, full information about the surgical procedure and recovery treatment were provided.

Between 2004 and 2012, there were no published articles in the area of implantable PV cells. During this period, implantable PV cells were discarded energy harvesters due to their size and due to the risk of infection from the connecting wires. However, thanks to advancements in integrated CMOS technology, there has been a renewed interest in using PV cells in 2012. In fact, Sahar Ayazian proposed a self-powered and fully integrated system, which embedded power-harvesting PV cells and sensor arrays in a 2.5 mm $\times$ 2.5 mm CMOS chip. They demonstrated successful power harvesting in the microwatt range for a device that was implanted 3 mm below the chicken skin. They tested the feasibility of using a CMOS p–n junction as a potential PV power harvester [11].

Furthermore, Haeberlin *et al.* successfully demonstrated a PV-driven pacemaker using thin-film silicon materials [39], which was tested in vivo for 40 days in 2015 [40]. The power density of their device was 0.95 mW/cm^2, as shown in Table 7.1. To scavenge more power, expensive semiconductor materials such as GaAs and GaInP, are needed [89]. The power density of GaInP/GaAs was 8.46 mW/cm^2 under AM1.5G condition [89]. The GaAs PV cell achieved higher power density of 44 mW/cm^2 by using NIR light source [55]. However, better encapsulation would be required due to the toxicity of these materials. In fact, A. Ahnood *et al.* demonstrated a diamond capsule for implantable PV cells [4], which can be used as not only a package but an optical window because of its high mechanical robustness, biocompatibility, and wide transmission spectrum [4].

Considering the comfortability of the patient, flexible PV cells are used in the implantable system by K. Song *et al.* in 2016 [90]. The PV cells were encapsulated with polydimethyl siloxane (PDMS). The device can successfully supply a commercial pacemaker in a rat with the power of 8 mW/cm^2. To test the biocompatibility of the device, the amount of arsenic was measured by using a mass spectrometer, and the 0.02 µg leakage by using emerging PDMS is lower than the daily intake from air breath (0.6 µg) and water (20 µg) by a person [89]. As the power output of the PV cell highly depends on the feature of covered skins, it is necessary to test the performance of the cells under human skins even if there are a lot of studies showing the viability in the animals. In this case, the performance of the PV cell under fresh and fixed human skins were measured in different location in 2017. The results (2.34 mW/cm^2 under inner arm, 2.21 mW/cm^2 under hand dorsum, 0.96 mW/cm^2 under forehead) show that the device under inner arm can provide most power output within same incident optical power [90]. Organic PV (OPV) cells were also used for retinal applications due to their sensitivity to NIR light [87]. The authors demonstrated both single-junction and tandem OPV cells based on a bulky heterojunction. They investigated the voltage and storage charging time of the stimulating electrode in a saltine solution via electrical pulses. Since the V_{oc} of tandem PV cells (1.31 V) was higher than that of single-junction PV cells (0.67 V), the tandem PV cell was able to provide a full charge per pulse stimulation window in NIR light conditions. In addition,

Table 7.1 A comparison of different types of implantable PV cells

Material	Source	P_{out} (mW/cm^2)	Pack.	Flex.	App.	Ref.
c-Si	NIR	9.35[b]	Diamond	No	Cochlear	[4]
c-Si	AM1.5G	6.17[a]	–	No	Sensor	[11]
c-Si	Halogen	18.87[b]	–	No	-	[18]
c-Si	Outdoor	4.94[b]	Silicone	No	Cardiac	[39]
c-Si	NIR	2.13[a]	–	No	Temp.	[48]
c-Si	NIR	2.21[b]	PLGA	Yes	LED	[61]
GaAs	NIR	44.12[b]	Silicone	Yes	Retinal	[55]
GAInP	AM1.5G	8.46[b]	PDMS	Yes	Cardiac	[89]
GaAs	LED	42.29[b]	–	No	Neural	[93]
α-Si	Halogen	0.95[b]	Silicone	No	Cadiac	[22]
α-Si	Outdoor	0.06[b]	Silicone	Yes	Temp.	[101]
Organic	NIR	3.84[b]	–	Yes	Retinal	[87]

Hints: (a) On-chip PV cell, (b) off-chip PV cell. "Pack." is Package, "Flex." is Flexibility, "App." is Application, and "Ref." is Reference.

the efficiency of the tandem PV cell was higher (5.6% in comparison to 5.3% in the single junction). For safety concerns, the light intensity in retinal applications was limited from 150 to 600 mW/cm^2. In this scenario, the tandem OPV with an active area from 2500 to 6250 μm^2 and electrode diameter of 35 μm can efficiently tune the charge per pulse, while the electrode of single-junction OPV was 60 μm, which limits the implantable resolution.

In summary, commercial PV cells were tested in vivo and used for implantable applications between 2000 and 2010. In effort to reduce the size of these cells (and increase the power density), CMOS technology was used after 2010 for fabricating both the on-chip PV cells and the power management circuitry. In the period between 2014 and 2017, researchers began using implantable PV cells to power real devices such as pacemakers. Moreover, encapsulation materials were used to protect device and ensure it is biocompatible. During this period, in vivo testing in animals and in vitro testing in human skin types were undertaken. Since 2018, emerging PV technologies such as organic PV cells have been used in implantable applications.

7.4 Power management in photovoltaic cells

In 2017, Z. Chen *et al.* presented an all-in-one system on a chip that operates under a solar energy harvester, along with a power management system. Best to our knowledge, this is the first paper to combine PV cell and on chip DC–DC converter. The following reviews are all based on system on-chip DC–DC converter with switched capacitor approach by using PV cell as the primary source of energy [18]. J. Zhao *et al.* in 2019 has presented the paper that focuses on varying input power rating in implantable solar energy harvester over the penetration of different ethnic group skin types [105]. In order to regulate the varying input and provide the reconfigurable gains and regulated output, the power management system presented by O.H. Kaung *et al.* (2018) is implemented. This single-input-dual-output (SIDO) power management design can produce simultaneous reconfigurable gain pairs and regulate the output with high conversion efficiencies [44,106]. X. Wu *et al.* in 2017 has proposed a switched capacitor DC–DC converter with discontinuous harvesting approach (DHA) to enable ultra-low power-energy harvesting system. The potential poor ambient conditions resulting the poor harvestable energy delivery from PV cell is considered and sustain the minimum of 40% efficiency [102].

In contrary to the conventional converters, in the process of stepping up voltage conversion at the load end, instead of continuously gathering charge from the energy source, the proposed (DHA) converter accumulates the charge slowly at an input capacitor and rapidly fire to the load capacitors. Therefore, the charge pump (CP) minimizes the efficiency lost for wide input range especially true for two conditions contribute toward poor efficiencies: (1) first, the low harvested input power pin of energy source in low ambient condition, for which operating frequency of charge pump needs to slow down due to slow energy delivery and increased leakages, (2) and second, increase in pin where charge pump needs to run fast to correspond the input power surge trigger constant overhead of clock power consumption and charge pump efficiency bound to driving strength of transistors. DHA is applicable during condition (1). In two-phase operation, DHA optimizes the trade-off between maximum power point tracking (MPPT) and charge-pump (CP) efficiencies to maximize end-to-end efficiency. Harvested voltage Vsol is deviated from VMPPT consequence in slight reduction in harvesting source efficiency and yet CP efficiency is improved by duty cycle through switches S2 and S3, which were enable through mode controller (MC). During first phase, Vsol attempts to increase above reference VMMPT from below VMPP and isolates the CP by disconnecting S2 and S3, so it sacrifices the MPPT efficiency and power gating of charge pump to reduce the leakage below 15pW.

Conventional CP, however, will hold Vsol at fixed VMPPT and enhance constant high leakage and low efficiency of CP in low ambient condition. When Vsol is adequate, a burst mode or quick charge transfer phase of start-up mode or phase 2 is operated in its peak efficiency at steady state. Then after short period, Vsol decreases and returns to harvesting mode [102]. Although the leakage can be minimized by reducing the power transistors size, this also limits the maximum harvestable energy. The end-to-end efficiency (of power management from energy harvesting sources) is defined by the multiplication of charge pump efficiency, which has many limitations due to leakages and other losses [23,84], and MPPT efficiency, which can be easily maintained for high input range. In this paper, we present a hybrid structure energy harvester that automatically modulates different duty cycle independently from charge pump operating frequency. X. Liu *et al.* in 2016 proposed the reconfigurable converter that allows access to fractional gain conversion ratios (CR) to optimize resolution of the harvestable voltage from energy sources as well as to minimize the inducing of a charge redistribution loss. This is important because to accommodate the environmental energy harvesters such as PV cell as a primary source of the power management system can lead to significant variations in harvested voltages due to environmental changes and these affect the voltage conversion efficiency (VCE) with fixed CR. The capacitive DC–DC topology is inherent to its structure and the number of stages, depending on the type of topology applied [60]. S. Modal *et al.* in 2016 proposed a PMS for solar energy harvesting. The novelty of the power conditioning design involves the simplicity of hardware, and the variation of ambient conditions are taken into consideration. In this context, PMS topology is categorized into three different conditions: (1) when there are sufficient ambient conditions that incur high harvestable energy to supply to desired load. In this scenario, a single DC–DC converter is employed to regulate the desire output power at the load and to store excess energy in a supercapacitor. This is done by comparing V_{DIV} and V_{REF1} – the charge transfer phase ($V_{DIV} < V_{REF1}$) involves transferring charge from C3 deliver to C_{load} and the store energy phase ($V_{DIV} > V_{REF1}$) from C3 to C_{STO} through switches M8 to M10. This stored energy is reused together with switching regulator when there is second condition (2) when there is low harvestable energy at poor ambient condition. This operation is governed by comparison of V_{DIV} and V_{REF2}. The charge transfer from C_{STO} to C_{load} occurs when $V_{DIV} > V_{REF2}$ through $M11$ as increase in drain potential of $M11$ is bigger than its source and turns the transistor on. (3) The like of varying ambient condition is also considered to maintain higher low-dropout (LDO) regulator (LDR) efficiency and addressed by introducing proper regulation at intermediate voltage between switching regulator and LDR [67]. Moreover, the same group has presented the implementation of architecture with measurement results of fabricated chip design [65] and analytical modeling in [66]. Recently in 2019, Y. Jiang and coworkers have proposed an algebraic series-parallel(ASP) topology for switched capacitor charge pump, which is known to be able to configure flexible fractional voltage conversions (VCRs) [51]. In conventional two-dimensional series–parallel (2DSP) [45,62], despite the flexible rational gain nature of VCRs, it often has large bottom-plate voltage swings incurred by the structure of circuit configuration that contribute toward parasitic loss. The improvement in this paper contributes more gain flexibility over rational gain ratios, which will not only improve operational output ranges but also help for system with large input variation range.

In conclusion, there are other literature which proposed ultra-low power-charge pumps that are worth exploring for a potentially candidate for the power management of the solar energy harvesting. For a start, the reconfigurable capacitor charge pump powered by thermal energy harvester for IoT applications presented by S. Yoon *et al.* [104] can operate at the ultra-low input-voltage range of 0.27–1 V with 64% efficiency and regulate toward 1 V output. Similarly, the proposed literature from H. Peng and group presented multiple start-up charge pumps that can be used for low input voltages [74]. Moreover, the proposed work of B. Mohammadi *et al.* achieved the highly energy efficient Dickson charge pump with clock blocker circuit with high efficiency of 89% [64].

The state of the art of the PV power management integrated circuits (ICs) is shown in Table 7.2.

Table 7.2   A state of the art for PV harvesting power management

Ref.	[18]	[44]	[102]	[60]	[65]	[51]
$In(V)$	0.25	1	0.25–0.65	0.45–3	0.33–0.405	0.25–1
$Out(V)$	852 m	1.49–2.8	3.8–4	3.3	1.01	1
P_{In}	1.44 μW	–	1.5 μW	–	1.19 mW	–
P_{Out}	1.65 μW	0.23 mW	–	50 mW	1.01 mW	1.2–20.4 mW
Cap	–	1 nF	1.5 nF	62.7 pF	500 pF	3 nF
VCR	3.5	3	–	4/3–8	3	1.25–5
Topology	*Dickson*	*SP*	*CP*	*CP*	*CP*	*ASP*
Area (mm^2)	0.24	–	2.72	4	994	0.54
freq. (Hz)	800 k	1.1 M	–	1 M	100 k	57 M
Stage	3	1	3	4	1	–
Eff.(%)	67	85.26	60	81	8.24	80

7.5   Design methodologies of implantable photovoltaic cells

In this section, the development of implantable PV cells is demonstrated. Different PV cell performance using different materials and technologies are also compared. Next, the fabrication process of PV cells and the encapsulation techniques are discussed. Here, it is important to emphasize that encapsulation involves protecting the PV cells as well as reducing the adverse side effects to humans. Optical losses due to tissue scattering effects are analyzed, and the optical properties of different skins are also be summarized.

Due to scattering losses from tissue, implantable PV cells are more effective in capturing light in the NIR range [17]. NIR light has problems with heating, though, which could make the wearer uncomfortable. Crystalline and noncrystalline materials can be utilized to fabricate PV cells, which fall into two general categories. In general, crystalline solar cells are less flexible, more expensive, and more efficient than noncrystalline alternatives [103]. In 2015, crystalline silicon materials made up about 93% of all PV cells; 24% of them were monocrystalline silicon and 69% were multicrystalline silicon [76]. Today, crystalline silicon makes up almost 89% of solar cells, amorphous silicon makes up 10%, and the remaining 5% are made of CdTe, diesel, copper indium, and gallium arsenide [41]. Table 7.1 compares the output power, flexibility, and the other features of distinct PV cells manufactured by different materials.

Silicon and gallium materials are frequently used in implantable applications due to their advanced technology [18,69,89]. For instance, gallium's greater bandgap and lower reverse saturation current make it particularly well suited for NIR light absorption, while a-Si is more adapted to visible light [86]. Additionally, because of the heating problems, Ga has a greater breakdown voltage and is less sensitive to heating 0.14 to 0.15 Ω resistance fluctuation based on the temperature coefficient from 0°C to 50°C [15]. For instance, the breakdown voltages of GaP and GaAs at 10^{15} cm^{-3} doping gradient are approximately 800 and 300 V, respectively, which is higher than that of silicon, which is 200 V [34]. Nevertheless, the high cost and the toxic nature are the main limitations to apply in implantable medical devices.

Biodegradable thin-film silicon PV cells have been researched in addition to wafer-based PV cells [86]. Microelectronics processing methods are required for the manufacture of the thin-film cell in comparison to the wafer-based cell cutting and sawing procedures [86]. Amorphous silicon (a-Si), microcrystalline silicon (μc-Si), and monocrystalline silicon (c-Si) are the most frequently utilized materials for thin-film PV cells. Crystalline Si thin-film cells are created throughout the production process using the transfer printing technique, whereas a-Si and polysilicon thin films are created using the Chemical Vapor Deposition (CVD) technique. OPV cells, on the other hand, have advantages because of their high flexibility, low cost, and simplicity of production [103]. They might make a promising contender for implantable applications if the efficiency of energy conversion can be increased, as well as the cell stability.

Currently, two categories of PV cell architectures are used in implantable applications: on-chip PV cells [11,17,18,48,49,55,61,68,69,89,90,93] and off-chip PV cells [4,22,39,101]. On chip PV cells are fabricated by using CMOS technology, - which applies the junction diode formed by N+ region, P+ region, N well, p well, deep N well, and p substrate [8]. For instance (shown in Figure 7.8(a)), one P+/N-well PV cell and one N+/P-sub PV cell were used as the PMOS source and NMOS source. Each PV cell was able to supply 0.3–0.4 V, and different configurations of PV cells can provide different voltage output [8,18]. To prevent photo-generated current from leaking into the CMOS circuit during operation, a metal layer is frequently used to shade the circuit during design and manufacture. Due to the integration of all the components into a single block, the on-chip PV cell takes up less space than an off-chip cell. However, the efficiency of on-chip PV cells is typically lower (20%), and it is challenging to create alternative PV cell configurations due to the negative voltage created by inverted junctions and the P substrate connected to ground [18].

In CMOS fabrication, double-well and triple-well technologies are mostly applied in forming the junction of on-chip PV cells, which includes different diffusion regions: P-type substrate/Deep N-type Well (P-sub/DNW), Deep N-type Well/P-type Well (DNW/PW), and P-type Well/Heavy N-type Region (PW/N+) [43]. The structure of typical CMOS on-chip PV cells is shown in Figure 7.8(b) and (c). Depending on their depth, CMOS-fabricated PV cells may produce varied amounts of energy. The performance of three different types of 180-nm TSMC PV cells was discussed utilizing a halogen light bulb with input power of 1.13 mW/cm^2 in Z. Chen *et al.* [18]. To be specific, the P-sub/DNW junction diode provides short-circuit current density, J_{sc}, of 433 pA/μm^2 as well as an open-circuit voltage, V_{oc}, of 0.55 V. Relatively, the J_{sc} and V_{oc} of the DNW/PW junction diode are 70 pA/μm^2 and 0.53 V, respectively, while the J_{sc} and V_{oc} of the PW/N+ junction diode are 74 pA/μm^2 and 0.53 V [18].

The fabrication of on-chip PV cells is based on the stacking of the junction diodes. Figure 7.8(b) and (c) demonstrates two stacking typologies to connect three served junction diodes. The topology in Figure 7.8(b) is proposed by M.K. Law *et al.*

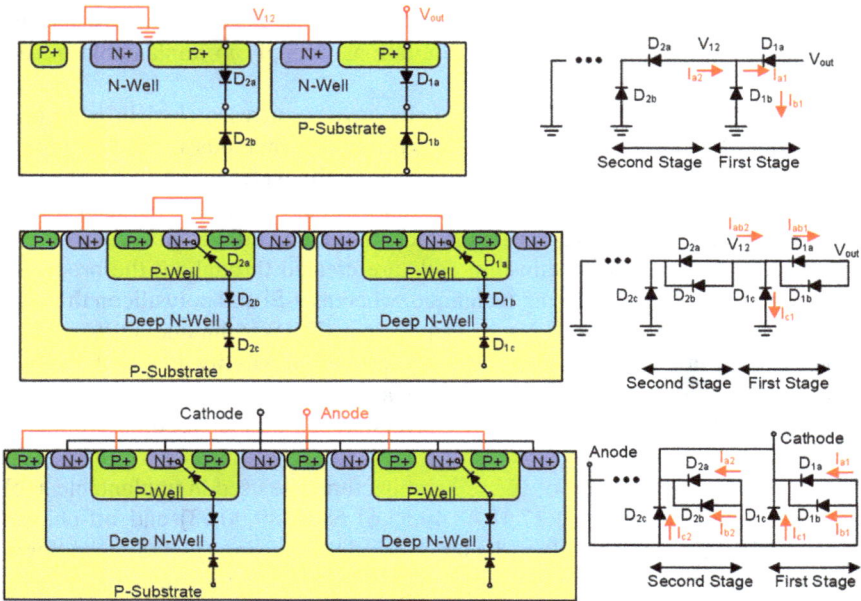

Figure 7.8   (a) The idea of fabricating on-chip PV cells in CMOS technology. Panel (b) shows stacking schematic and equivalent circuit of double-well 0.35-μm CMOS technology by M.K. Law et al. [56]. Panel (c) shows stacking schematic and equivalent circuit of triple-well 0.25-μm CMOS technology by G. Hong et al. [43].

[56], which applied double-well 0.35-μm CMOS technology. The power conversion efficiency of the stacking junction diodes is evaluated by [43]

$$K > 1 + \alpha \tag{7.11}$$

$$\eta = \eta_0 \times N \frac{K - 1}{K^N - 1} \tag{7.12}$$

where K is the area to volume ratio, α is the leakage photocurrent ratio $\alpha = I_{bN}/I_{aN}$), η is the overall efficiency while η_0 is the first-stage efficiency, and N is the number of stages. A high K value affects efficiency; thus, it is best to keep it as low as possible. However, K is constrained by $1 + \alpha$. In this situation, minimizing alpha would result in a smaller K. I_{bN} should be reduced to a minimum and I_{aN} should be increased [43,56]. In the triple-well technology, the current I_{abN} is increased by connecting a DNW/PW (D_{Nb}) and PW/N+ (D_{Na}) diode in parallel. The amount of optics reaching the P-sub/DNW D_{Nc} decreases when the D_{Nb} is inserted deeper, which lowers I_{cN}. The first and second methods have the advantage of boosting PV cell output voltage, which makes it easier to support MOSFETs in the energy harvesting circuits on the same chip. Additionally, one P-sub/DNW diode in the final stage is bypassed, meaning that not all of the diodes are being utilized [43]. Compared with the conventional use of monocrystalline silicon and the limited stacking topologies of CMOS technology, off-chip PV cells can be based on a variety of materials (inorganic and organic) and stacking topologies. For instance, heterojunction PV cells were previously fabricated using GaInP/GaAs materials in a 2×7 array to drive a load [89].

7.6   Finite-element method in implantable photovoltaic cell fabrication

The literature provides a wealth of tools for numerically simulating PV cells [6]. The earliest simulations involved solving a set of "continuum" partial differential equations (PDEs), commonly known as the semiconductor equations [94]. Thanks to early efforts at Bell Labs in the 1960s, computer simulations demonstrated that PV cell efficiencies can reach 19% [27,28]. Improvements in computers enabled these simulations to be performed on personal computers (PCs). Gray and Basore from Purdue University were best known for initiating these modeling efforts during the 1980s and early 1990s with their one-dimensional (1D) simulations program called PC1D, which solved the semiconductor equations using the finite-element method (FEM) [13]. Later, a more advanced simulation tool called "A Device Emulation Program and Toolbox" (ADEPT) was developed, which enabled two-dimensional (2D) and three-dimensional (3D) simulations to be performed on crystalline-based PV cells [36]. Currently, Sentaurus Device is the latest commercial tool for simulating PV cells in 2D and 3D from Synopsis [5]. For example, the Atwater group in Caltech used this software to characterize the performance of nanowire silicon solar cells [52].

As previously mentioned, crystalline silicon-based solar cells dominate the PV market [5]. Further penetration of solar energy solutions in the market relies on developing lower cost cells and improving their efficiency. For example, increasing the efficiency of solar cells requires innovations in light trapping and in creating antireflection coatings [1,10]. However, enhancing the performance of crystalline solar cells is a nontrivial and complex task, as the performance relies on ambient conditions, recombination within the semiconductor, and various parasitic losses [5].

To predict the performance of solar cells, the FEM has often been used for solving the semiconductor equations describing the electrical carrier properties of solar cells in 1D, 2D, or 3D [5]. In the 1D and 2D models, carrier generation and optical absorption are rarely analyzed thoroughly, and only low-dimensional approximations are applied [59]. Since 1D FEM simulations often use the Beer–Lambert method to calculate carrier transport, 1D tools can be used to simulate basic cells without features in the direction parallel to the p–n junction [71]. Compared with the optical approximations in 1D modeling, the EM response is calculated using the Jones matrix method in 2D and 3D, which provides more accurate optical analysis [59].

According to the literature, 3D software tools enable accurate simulations that agree with experiments, as parasitic losses can be included. Some of these parasitic losses can be considered using the solar cell's equivalent electrical circuit and its subsequent current–voltage (IV) response equation [59]. These losses can be defined in terms of the equivalent "series" and "shunt" resistances.

Nevertheless, the majority of these software tools require expensive licensing fees. Thus, I review the range of free software tools for simulating PV cells. I discuss the merits and limitations of these tools. Moreover, I provide a step-by-step demonstration of how to simulate a simple p-i-n solar cell using a commercial multiphysics FEM tool. Comparisons and analyses between the simulation tools are provided.

7.6.1   *Overview of finite-element-method modeling software*

The earliest PV FEM modeling tool for solar cells was PC1D, as shown from the timeline in Figure 7.9. Moreover, a list of free PV modeling tools is shown in Table 7.3.

For example, PC1D and PC3D are two noncommercial tools for 1D and 3D simulations, respectively. In the literature, PC1D was previously used to optimize the

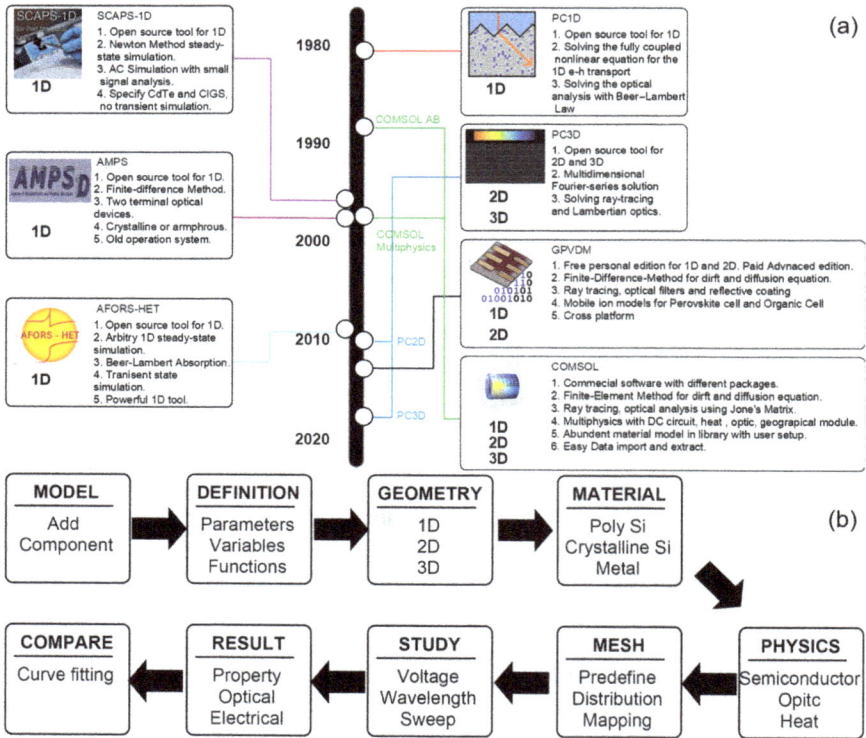

Figure 7.9   *(a) Timeline showing the development of four different PV cell simulation tools: PC1D, PC3D, GPVDM, and COMSOL. A comparison between their main features is also shown. (b) Diagram showing the modeling workflow in COMSOL FEM simulation. Briefly, the tool enables solar cell performance to be predicted by defining the material parameters, geometry, meshing model, and other numerical simulation properties.*

Table 7.3   Comparison between the device modeling software

Software	Dimension	Mathematical solver	Focused theme
COMSOL	1D/2D/3D	FEM	General physical theme
GPVDM	1D	FDM	C-Si/A-Si/CIGS device
PC1D	1D	Newton method	General PV device
PC3D	1D/3D	3DFEM	General PV device
AMPS	1D	FDM	C-Si/A-Si device
SCAPS-1D	1D	FDM	Poly and thin-film CdTe and CIGS
AFORS-HET	1D	FDM	Arbitrary semiconductor

efficiency of a monocrystalline silicon solar cell with an efficiency of 20.35% [42]. Similarly, due to its 3D modeling capabilities, PC3D was used to accurately simulate the optical properties of light interacting with the PV cell's textured surface as well as its top interdigitated electrode structure [14].

The General-Purpose Photovoltaic Device Model (GPVDM) is another semiconductor simulator that enables users to simulate emerging PV cells, such as perovskite cells. For example, it was used to investigate the performance of perovskite PV cells when the active layer thickness and temperature were varied [2]. This tool also provides a 3D graphical representation of a PV cell, but only solves the semiconductor equations in 1D. It has an intuitive output interface that enables users to obtain both electrical and optical PV data, such as the photo absorption rate and the generation rate at different regions within the thickness of the device.

Moreover, Analysis of Microelectronic and Photonic Structures (AMPS) is another free 1D multipurpose simulations tool, which has been used to investigate the effect of varying the absorber layer thickness in CIGS PV cells for a range from 300 to 3000 nm [79]. It has a library of monocrystalline, polycrystalline, and amorphous materials. It can produce Current Density versus Voltage (JV) characteristic curves as well as quantum efficiency. However, the tool can only be installed on the Win XP platform, as its developers do not support newer operating systems or platforms.

Furthermore, Solar Cell Capacitance Simulator 1D (SCAPS-1D) has been used for modeling polycrystalline Si PV cells [16], perovskite-based PV cells [88], and thin-film PV cells based on CdTe [46] and CIGS [70].

Automat for Simulation of Hetero-structures (AFORS-HET) is another free simulator used to model an arbitrary 1D sequence semiconductor interface [29]. It has been used to simulate heterojunction PV cells [31]. It can produce both electrical and optical PV data. AFORS-HET can also simulate materials with graded bandgaps as well as carrier transport across the various interfaces [80]. However, it is not capable of simulating high-dimensional models of solar cells.

The performance of these noncommercial tools is compared with COMSOL, which is a commercial software program that solves the semiconductor PDEs in 1D, 2D, and 3D. In addition to its higher dimensional modeling capabilities, COMSOL has an intuitive and easy-to-use graphical user interface that enables users to simulate different domains and boundary conditions. It has been previously used for simulating the performance of implantable PV cells [105]. Compared with COM-SOL, these noncommercial tools have a smaller file size and are capable of solving the semiconductor equations in 1D and 3D.

The following simulation platforms were used during our investigations: COM-SOL Multiphysics 5.4.0.388 with the Semiconductor and Waveoptic modules, PC1D v5.9, PC3D v1.7, and AFORS-HET v2.5. The simulations were performed on a computer with an Intel Core i5-6300U processor (2.40 GHz), 4 GB RAM, and a Windows 10 64-bit operating system.

7.6.2   Definition and setup

Initially, the parameters, variables, and functions were set up in different scenarios, light intensities, and device properties. The parameters, functions, and variables for the FEM simulation are summarized in Table 7.4. At first, the electrical properties of silicon were defined, which included the intrinsic doping concentration [57], relative permeability [57], electron affinity [57], bandgap [57], density of states [21], and carrier mobility [7].

The density of states and carrier mobility functions are shown in Figure 7.10(a) and (b). The recombination losses will be determined by the recombination coefficients, such as the Auger coefficient, Radiative coefficient, and SRH coefficient, which are shown in Figure 7.10(c) and (d), as well as in Table 7.4.

Table 7.4   Summary of device simulation parameters

Parameters	Function of value	Description
Equation iteratively solved	Poisson's Equation	X. Li *et al.* [59]
Free carrier statistics	Femi-Dirac	P.P. Altermatt *et al.* [7]
Temperature	300 K	P.P. Altermatt *et al.* [5]
Intrinsic doping density	1×10^{10} cm^{-3}	M. Levinshtein *et al.* [57]
Relative permittivity	11.7	M. M. Levinshtein *et al.* [57]
Bandgap	1.12 eV	M. M. Levinshtein *et al.* [57]
Electron affinity	4.05 eV	M. M. Levinshtein *et al.* [57]
Density of state	Couderc model (Figure 7.10(a))	Couderc *et al.* [21]
Carrier mobility (μ_p, μ_n)	parameterization (Figure 7.10(b))	P.P. Altermatt *et al.* [5]
Reconbination coefficients		
Auger (C_p, C_n)	Dziewior and Schmid (Figure 7.10(c))	P.P Altermatt *et al.* [5]
Radiative (B)	Nguyen model (Figure 7.10(d))	Nguyen *et al.* [72]
Optical parameter		
Refractive index	Silicon (Figure 7.10(e))	Aspnes *et al.* [9]
Optical spectrum	AM1.5G (Figure 7.10(f))	Nguyen *et al.* [72]

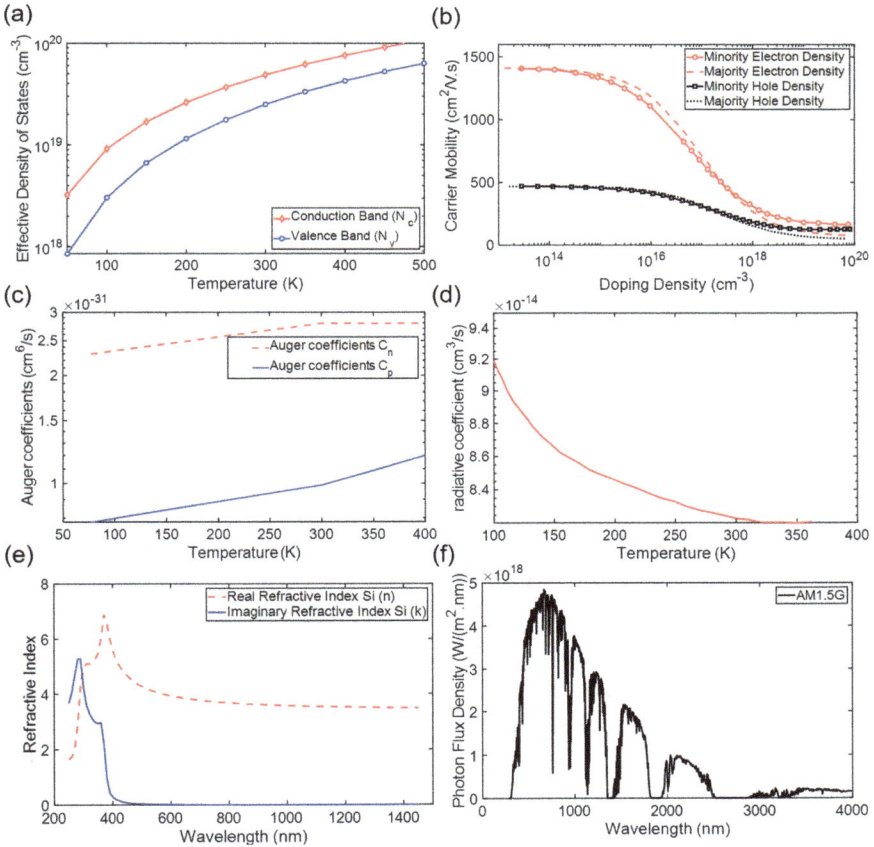

Figure 7.10   (a) Density of states model from Couderc et al. [21]. (b) Carrier mobility parameterization from Klaassen et al. [5]. (c) Auger coefficients of Dziewior and Schmid, which were obtained from P. Altermatt et al. [5]. (d) Radiative recombination coefficient from Nguyen et al. [72]. (e) Refractive index of crystalline silicon from Aspnes et al. [9]. (f) Photon flux density of AM1.5G spectrum from Nguyen et al. [72].

The absorption of light and generation of an electron–hole pair are fundamental to PV cell operation. In this process, the energy of a photon is initially converted to electrical energy through the creation of an electron–hole pair [6]. The refractive index of silicon was used to analyze the amount of light absorbed or penetrated in the PV cell, as shown in Figure 7.10(e).

Moreover, for simulating the performance of PV cells, the AM1.5G global irradiance spectrum was used, where a power density of 1000 W/cm^2 was assumed. The photon flux spectrum shown in Figure 7.10(f) was used to set up the generation rate [72]. Unless otherwise stated, an ambient temperature of 300 K was used [57].

Table 7.5   Summary of device geometry and doping profile

Parameter	[97]	[82]
t	800 nm	150 μm
N_{P+}	1×10^{20} cm^{-3}	1×10^{19} cm^{-3}
N_{Base}	1×10^{16} cm^{-3}	3.5×10^{15} cm^{-3}
N_{N+}	1×10^{20} cm^{-3}	1×10^{17} cm^{-3}
L_{p+}	50 nm	1 μm
L_{P-}	500 nm	148.45 μm
L_{N+}	250 nm	250 nm
I_{rad}	1000 W/m^2	1000 W/m^2
λ	550 nm	550 nm
$\lambda Sweep$	400 − 1000 nm	400–1000 nm

7.6.3   Geometry design

Simulating an ideal p-i-n junction diode requires a uniform doping profile in each region. In our simulations, I have defined the different layers as heavy n-type doped (N+), light p-type doped (P−), and heavy p-type doped (P+). The length of these layers was L_{p+}, L_{N+}, and L_{P-}. The device properties and parameters are summarized in Table 7.5. I used these parameters to compare the performance of our simulations tools with experimental data from two different sources [97] and [82].

In our simulations, light penetrates from the bottom of the cell with an incident angle of zero, as shown in Figure 7.11. Therefore, the front surface of the cell was located at $y = 0$. Moreover, two metal contacts were added to the front and rear surfaces of the PV cells.

7.6.4   Material alignment

Polycrystalline silicon was used as the device material, with the properties shown in Figure 7.11. An "air" layer was defined on the top and bottom sides of the PV cell with a refractive index of ($N_{Air} = 1.0003$) [20]. Since the doping concentration is different in each layer, parameters such as carrier mobility and recombination coefficient are different. In this case, the materials were set up differently and aligned to specific layers (N+, P-, and P+) with different doping concentrations N_{p+}, N_{N+}, and N_{P-}.

7.6.5   Physical model description

To analyze carrier transport across the heterojunction interfaces, the semiconductor module enables important semiconductor function definition, such as doping, generation, recombination, trap density, and space charge density. On the other hand, PC1D, GPVDM, and AFORS-HET enable users to customize the doping concentration and diffusion length, while PC3D enables users to define the doping concentration by changing the sheet resistance of each layer.

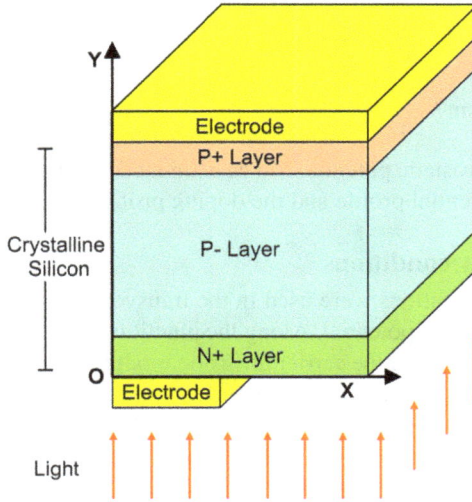

Figure 7.11   *Structure and architecture of the experimental PV devices in [82,97].*
For comparison, I simulated the same structure using COMSOL as
well as the noncommercial software programs.

7.6.6   *Optical calculations*

To determine the optical characteristics in 2D, the 2 × 2 Jones Matrix method was
used [9]:

$$\begin{bmatrix} E_{Ra} \\ B_{Ra} \end{bmatrix} = \sum_{i=1}^{m} \begin{bmatrix} \cos\phi_i & \left(\frac{j}{\eta_i}\right)\sin\phi_i \\ j\eta_i\sin\phi_i & \cos\phi_t \end{bmatrix} \begin{bmatrix} 1 \\ \eta_m \end{bmatrix} \qquad (7.13)$$

where δ_i is the wave phase shift ($\delta_i = 2\pi N_i d_i \cos\theta i / \lambda$) in the ith layer, N_i is the refractive index, d_i is the thickness of the ith layer, and η_i is the pseudo index in i^{th} layer
($\eta_i = N_i\cos\theta_i$). E_{Ra} or B_{Ra} is the ratio between the electric and magnetic fields of the
transmitted light and incident light. M is the total number of layers. The reflectance
(R), absorptance (A), and transmittance ($T = 1 - A - R$) of light can be determined
using (7.1)–(7.3).

7.6.6.1   Initial condition setup

Initially, I applied the neutral charge condition to setup the doping profile and ohmic
contact characteristics for estimating the initial electrostatic potential. The initial doping profiles for p-type (p_{init}) and n-type (n_{init}) doping were defined by applying the
neutral charge condition, where $n_{init} - p_{init} - C = 0$ and $n_{init}p_{init} = n_i^2$. These can be
described in terms of

$$n_{init} = \frac{1}{2}\left(\sqrt{C^2 + 4n_i^2} + C\right) \qquad (7.14)$$

$$p_{init} = \frac{1}{2} \left(\sqrt{C^2 + 4n_i^2} - C \right) \tag{7.15}$$

$$\Phi_{init} = \frac{kT}{q} \arcsin h \frac{C}{2n_i} \tag{7.16}$$

This initial electrostatic potential can be estimated for ohmic contacts. It can also predict the initial potential profile and the doping profile in a homojunction.

7.6.6.2   Boundary conditions

Periodic boundary conditions were used in the transverse directions for both carrier diffusion and electrostatic potential. Along the junction interface, surface recombination conditions were used for the carrier diffusion modules. For the Poisson equations under forward bias, the cathode was grounded and unchanged, but the voltage at the anode was varied from an initial value of zero to the forward bias voltage. In COMSOL and AFORS-HET, the boundary conditions can be customized.

7.6.7   *Electrical output characteristic*

First, the electron current density (J_n), hole current density (J_p), and short-circuit current density (J_{sc}) can be determined using

$$\mathbf{J}_n = -q\mu_n n \nabla \psi - qD_n \nabla n \tag{7.17}$$

$$J_p = -q\mu_p p \nabla \psi - qD_p \nabla p \tag{7.18}$$

$$J_{sc} = \iiint |J_n(x, y, \lambda) + J_p(x, y, \lambda)| \, dx \cdot dy \cdot d\lambda \tag{7.19}$$

Considering the parasitic resistances, the output current density of the PV cell can be obtained using

$$J(V) = -J_0 e^{-\frac{V}{V_T}} + J_{sc} - \frac{V + J(V)R_s}{R_{sh}} \tag{7.20}$$

where R_s and R_{sh} are the series and shunt parasitic resistances, and V_T is the thermal voltage ($V_T = kT/q$). Since our simulations model is in 2D, R_s can be neglected and R_{sh} can be calculated using the current–voltage (IV) characteristics of the device. The external quantum efficiency (EQE) can be determined using

$$EQE = \frac{hcJ_{sc}}{q\lambda I_{rad}} \tag{7.21}$$

where h is the Plank constant, c is the speed of light, and I_{rad} is the irradiance under AM 1.5G.

7.6.8   *Semiconductor and simulation setup*

In PC1D, GPVDM, and AFORS-HET, the doping concentration and diffusion length in each semiconductor layer need to be defined. However, in PC3D the setting is different, as the relative doping concentrations need to be converted into a sheet resistance. The doping profiles were set according to the parameters given in Table 7.5.

In the PC1D, PC3D, GPVDM, and AFORS-HET, only the surface recombination can be customized, while COMSOL enables users to define different types of recombination. Moreover, PC1D, PC3D, GPVDM, and AFORS-HET enable users to change the irradiance settings as well as the surface reflection.

In COMSOL, two types of doping profiles can be defined, which are "gradient" and "uniform." To ensure consistency with the noncommercial tools, I have used the "uniform" doping in PC1D, PC3D, GPVDM, and AFORS-HET as well as COMSOL. Then, the generation function was set to excite the electron–hole pairs in the device. Furthermore, in the COMSOL program three types of recombination were defined by invoking Equation (7.8).

Since the doping concentration is high in the heavily doped regions (N+ and P+), the bandgap narrowing functionality in COMSOL was used. Next, two metal contacts (anode and cathode) were defined to complete the PV cell architecture. Moreover, when selecting a "Terminal Type," there were four available options in COMSOL, which were Charge, Voltage, Circuit, or Terminated. In this case, I selected "Voltage." Finally, for "Contact Type," I selected "Ideal Ohmic." To complete our analysis of these PV cells, I investigated their performance concerning voltage as well as forward biasing voltage. I therefore varied the voltage between 0 and 0.7 V, whereas the wavelength range was 400 nm $< \lambda <$ 1000 nm.

7.6.9   Comparison results of different device modeling tools

I compared our simulation results with experimental data from [82,97,99]. The PV cell properties are provided in Tables 7.1 and 7.5. As previously mentioned, the modeled solar cell consists of five stacked layers: air, a layer of N+ silicon, an intrinsic layer, P+ layer, and another layer of air. A comparison between the JV curves for the simulation programs is shown in Figure 7.12. Despite using the parameters, the simulation tools provided different results. I discuss which tools provided better agreement with experimental data.

Clearly, the simulation results from COMSOL 2D and PC1D are closely matched to experimental data. However, results from COMSOL 1D are different from experimental data. In fact, there is almost a 2 mA/cm^2 difference in J_{sc}), as shown in Figure 7.12(a). I believe that this could be attributed to the optical approximation invoked by using the Beer–Lambert Law. Despite the difference in J_{sc}, the V_{oc} data from PC1D as well as COMSOL 1D and 2D all agree with experimental data (0.64 V).

Based on these simulation results, PC1D can be regarded as a reliable software tool that agrees with experiments. This is perhaps why it has been regularly used by the PV industry for decades. With COMSOL 1D and 2D, there are plenty of library resources and multiphysics modules that can facilitate the modeling of different materials and device architectures. Furthermore, numerical solutions to the PV equations were achieved in less than 5 s with COMSOL 1D, while the 2D COMSOL simulations needed 10–30 s.

With AFORS-HET, our PV cell modeling achieved $J_{sc} = 11$ mA/cm^2 and $V_{oc} = 0.6$ V, which compares favorably with experimental as well as the COMSOL

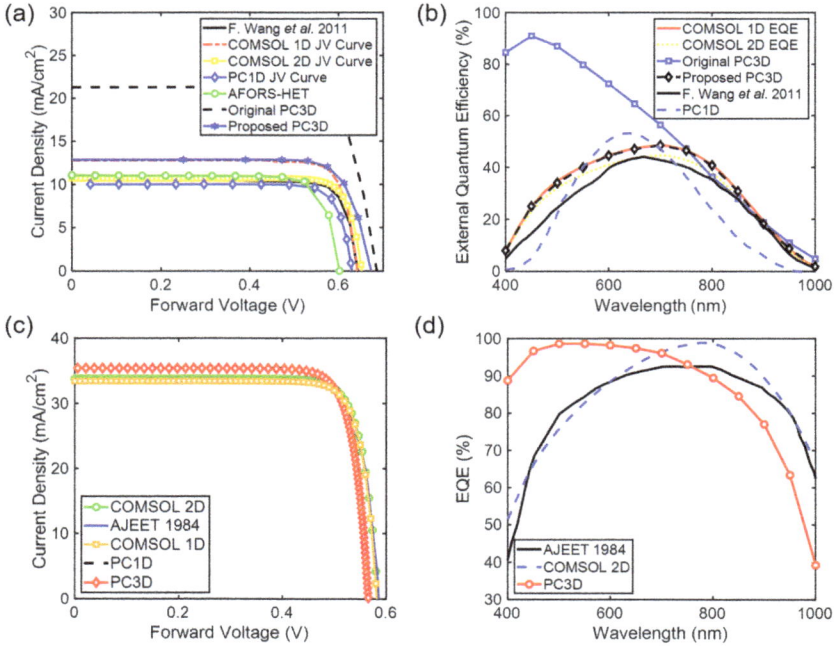

Figure 7.12   (a) *Comparison between the simulation results from PC1D, PC3D (original and proposed), COMSOL, and AFORS-HET and the experimental data from F. Wang et al. [97]. (b) EQE results from PC1D, PC3D, COMSOL 1D, and COMSOL 2D and the experimental data from F. Wang et al. [97]. (c) Comparison between simulation results from PC1D, PC3D, and COMSOL 2D and the experimental data from A. Rohatgi et al. [82]. (d) EQE results from PC1D, PC3D, and COMSOL 2D and the experimental data from A. Rohatgi et al. [82].*

data. In fact, AFORS-HET is useful as it allows the user to define different interface materials and boundary conditions. These functionalities are not available in the other surveyed 1D simulation tools.

As shown from Figure 7.12(a), PC3D provides slightly different results. Initially, $J_{sc} = 21.3$ mA/cm^2 and $V_{oc} = 0.69$ V. This large difference in results was due to using the default spectral transmission parameters in PC3D, where no reflectance was assumed. However more accurate results can be obtained by modifying the optical parameters. As previously mentioned, the front spectral reflectance was set to 70% [47] and the back reflectance was set to 50% [91], which enabled the EQE results to match those from COMSOL 1D, as shown in Figure 7.12(b). Thus, an improved curve with $J_{sc} = 12.9$ mA/cm^2 and $V_{oc} = 0.67$ V was obtained.

With PC3D, the software can only simulate devices with thicknesses exceeding 1 μm. This is an important limitation, especially for wearable PV devices, where thin-film materials are now attracting interest. For the sake of comparison, the experimental device in [97] is 20% smaller than this limit, whereby the thickness of the PV cell was 800 nm. Naturally, since the simulated PV cell is thicker than the experimental cell in [97], J_{sc} =12.9 mA/cm^2, which is larger than the J_{sc} reported in [97] (10.5 mA/cm^2). Consequently, there is a -22.9% difference between these experimental and simulation J_{sc} results.

Due to this limitation with PC3D, I have also investigated a thicker PV cell with a thickness of 150 μm), as reported in [82]. From our simulation results in PC3D, $J_{sc} = 35.5$ mA/cm^2, which is much closer to the experimental results mentioned in [82] ($J_{sc} = 36.1$ mA/cm^2), as can be verified from the results in Figure 7.12(c). Therefore, the difference between the experimental and simulation results is now only 1.7%.

As for the EQE results shown in Figure 7.12(b) and (d), COMSOL shows better agreement with experimental data compared with PC1D and PC3D. The mismatch in results is mainly due to the material and optical properties setup in these programs. For example, in COMSOL three types of recombination can be defined, which are Radiative, Auger, and Schockley–Read–Hall recombination. On the other hand, in PC1D and PC3D only surface and bulky recombination can be defined. Furthermore, in PC3D, the EQE is usually calculated using the previously mentioned default optical parameters. Including both the front and back spectral reflectance leads to better agreement with experimental data, as shown from the "Proposed PC3D" curve results in Figure 7.12. Consequently, this proves that the PC3D simulation results can be setup to show good agreement with experimental data.

Clearly, COMSOL is a software tool with a vast materials library, great curve fitting features, and a multiphysics environment, which produces results that agree with experimental data. It can therefore be used to accurately predict the performance of PV cells for wearable applications, particularly when such cells are exposed to nonuniform lighting conditions. In that case, the 3D PV cell model can be used to investigate the change in irradiance along the horizontal and vertical axes. Second, wearable applications require highly flexible devices for improved user comfort. In 3D modeling, it is easier to analyze the stress and pressure inside the device. Eventually, the PV cell semiconductor model in 3D COMSOL can be easily integrated with other multiphysics modules to investigate the impact of heat, pressure, and other environmental effects. In this case, PC3D is an alternative and free option that can perform many of COMSOL's simulations, provided that devices are larger than 1 μm. This is also especially true if the device properties library is complete, with an ability to modify the optical properties. I can also obtain more accurate results if multiphysics simulations can be involved, where the effects of temperature and stress can be investigated.

7.7　Device design concerns in implantable photovoltaic cells

In contrast to conventional PV designs, implantable PV cells have to consider additional optical losses associated with packaging materials and biological tissues. Since the standard PV cell is designed mostly for light absorption and reduced reflection losses at their incident surface, in the case of implantable PV cells, this introduces a new challenge of reduced transmission due to intervening layers. These are encapsulating materials that are designed for biocompatibility, and the innumerable tissue layers that normally attenuate light while it penetrates the body. This means that the development of implantable PV cells needs not only high efficiency in energy conversion but also innovative solutions to minimize the optical losses due to inherent optical properties of the body and protective packaging.

7.7.1 Package and encapsulation

To prevent the electrical circuitry from corroding, encapsulation is required. The materials used for conventional encapsulation are ceramics, glass, and metals such as titanium and ceramic packing. The majority of the titanium shell in the pacemaker and other implants packaged similarly is accessible to bodily fluids. If all exposed metal surfaces, including tracks, have the same potential, then this is okay. Metal is occasionally used in glass breakthroughs for such implants, however, this is becoming prevalent [40].

Materials that are extremely biocompatible include metals, glass, and ceramics. They are rigid and have a low water vapor permeability [85]. Despite their advantages in harsh environments, many conventional materials are incompatible with the CMOS production process [3,85]. Consequently, research has been done into new encapsulating materials such as SiO_2, SiC, Al_2O_3, diamond. The other flexible material includes organic silicone polymers, polyimide, PDMS, liquid crystals, and SU-8 is also involved [3].

Three considerations, including optical characteristics, biocompatibility, flexibility, and lifetime, can have a big impact on the choice of innovative materials for encasing PV-driven implanted devices. Figures 7.13(a),(b) and 7.14(a),(b) show the difference in the physical properties of materials which is normally applied in package and encapsulation manufacture. The materials include:

1. **Solid**: Crystalline SiO_2 [33], Diamond [75], Fused Silicon Dioxide (Fused SiO_2) [77], Crystalline Alumina (α-Al_2O_3) [78], Thin-film Alumina [78], thin-film SiO_2 [81], Hafnium Oxide (HfO_2) [100].
2. **Flexible**: Liquid crystal polymer (LCP) [58], Parylene [73], Silicone (PDMS) [78], Polymide [95].

It is significant to note that these refractive indices in the visible and NIR range are very close to 1.5, which is the same as the refractive index of skin [12]. Additionally, the materials' extinction coefficients in the visible and NIR ranges are nearly zero, resulting in low light absorption and great optical transparency. These mean time to failures (MTTFs) of the materials are determined by exposing them to a certain temperature, and the long-term reliability is verified in accelerated aging studies by immersing the samples in a saline solution. The MTTF values (reaction rate in saline solution) and other features of the materials can be used to determine how long these materials last in the human body [3]. The Young's modulus variation for various materials is shown in Figure 7.14(b). A high Young's modulus typically indicates that the material is stiff and has little flexibility.

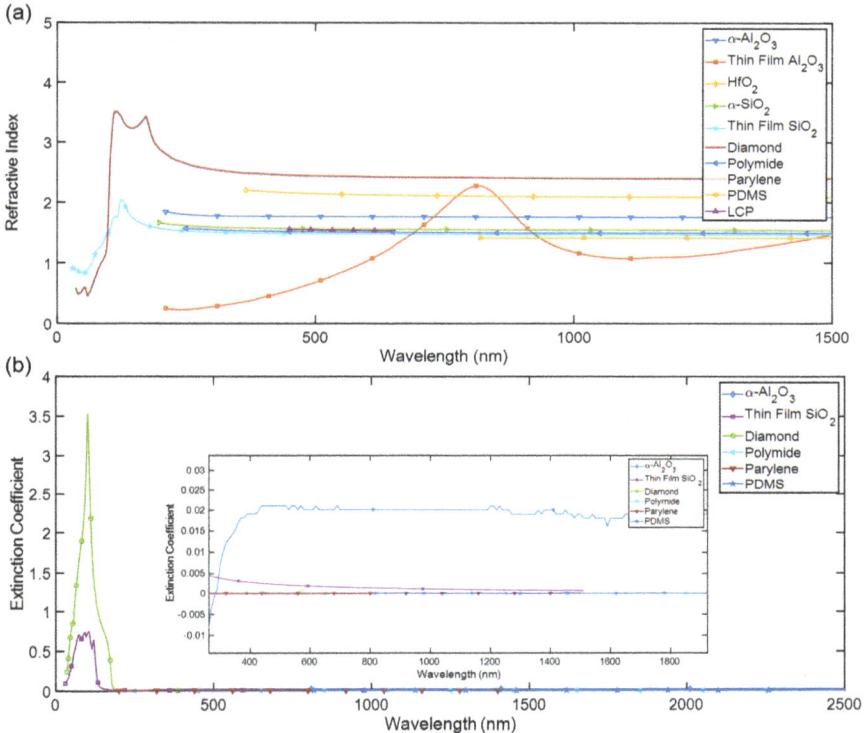

Figure 7.13 *Variation in the encapsulation material properties: (a) refractive index as a function of wavelength; (b) extinction coefficient as a function of wavelength*

The fabrication methods and physical characteristics of the encapsulating materials are shown in Table 7.6. According to the manufacturing technology, SiO_2 combines thermal bonding and CVD, whereas Al_2O_3 is often deposited by ALD and CVD. However, spin coating and heat bonding are typically used to generate organic encapsulation, such as polymide, silicone, PDMS, and LCP (liquid crystal polymer). Despite the presence of Parylene C, complicated surface topography with fissures and sharp edges typically requires the application of CVD.

Alumina (Al_2O_3) is frequently used as the feedthrough with a titanium package because it is regarded as a good encapsulation material with high biocompatibility [3]. Typically, silica is used in anodic bonding and laser welding. However, this encapsulation material has a problem with minimization [3]. A conformal coating of HfO_2 is frequently applied to nanoscale electronics using the atomic layer deposition (ALD) method. It offers excellent mechanical, chemical, and thermal stability [3,19]. Due to its mechanical rigidity, wear resistance, and chemical resistance [3,24], diamond is an emerging material for biomedical encapsulation. High pressure and temperature are necessary for the encapsulation process. They are perfect for implantable PV cells due

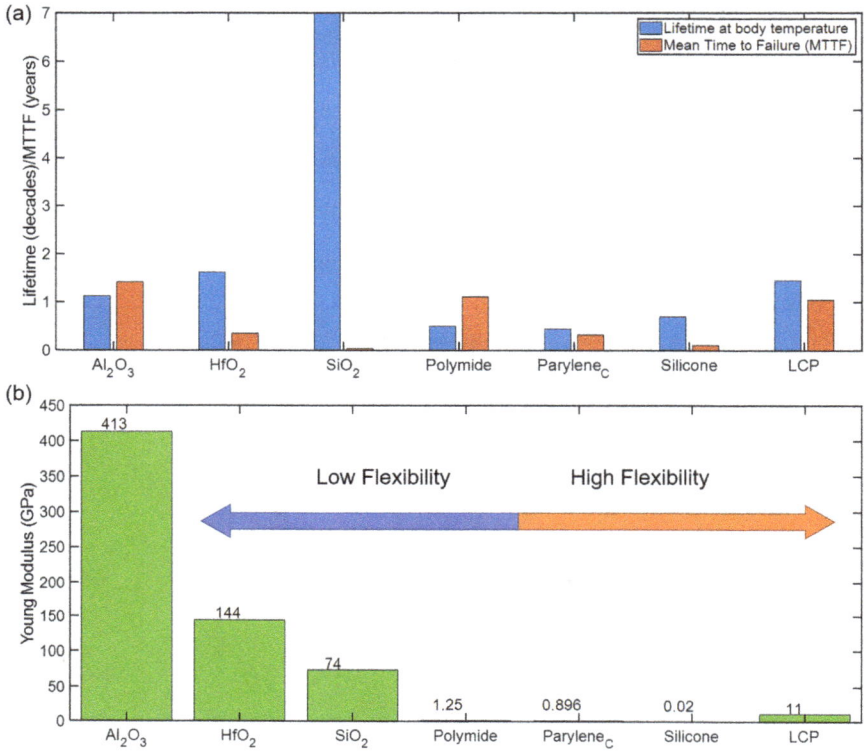

Figure 7.14   Variation in the encapsulation material properties: (a) lifetime in body temperature and mean time to failure, as well as (b) Young's modulus

Table 7.6   Comparison of key factors for different types of encapsulating materials

Material	Tech.	Thick (μm)	Thermal[a]	Moisture (%)[b]	Flex.[c]	Failure (nA)[d]
Al₂O₃	ALD+CVD	6.052	$4 \times 10^{4°}C^{-1}$	4.75[e]	No	1
HfO₂	ALD	0.1	–	–	No	1
SiO₂	Thermal	0.1–1	$4.5 \times 10^{4°}C^{-1}$	6.04[e]	No	–
Diamond	CVD	300	–	–	No	–
Polyimide	Spin Coating	10	$40\ ppm\ °C^{-1}$	3	Yes	1000
Parylene C	CVD	6–40	$35\ ppm°C^{-1}$	–	Yes	1
Silicone	CVD	40	$340\ ppm°C^{-1}$	–	Yes	–
PDMS	Spin Coating	150	–	–	Yes	–

(a) Thermal expansion coefficients.
(b) Moisture absorption.
(c) Flexibility.
(d) Failure criteria.
(e) Moisture absorption at 30°C.

to their optical characteristics [4]. The majority of micro-electro-mechanical systems (MEMS) procedures can work with polyimide (photosensitive polyimide in photolithography and polymide adhesion in bonding). Since parylene is flexible, inert, and with great optical transparency, it is advantageous in thin conformal layers [3]. Due to its transparency and broad refractive index range (1.38–1.58) [53], silicone is frequently used as an encapsulant for optical devices such as LEDs, photodetectors, and PV cells. In addition, silicone meets the typical requirements of the healthcare and electronics industries due to its low dielectric constant (2.68 at 100k Hz), high thermal conductivity (0.4–1.34), and low resistivity ($2.9 \times 10^{14} \, \Omega \cdot m$) [53]. The comfort layer is well protected for encapsulation purposes by silicone rubber, which can also be combined with ceramic material to create an impermeable enclosure. The following are some benefits of silicone:

1. The outer layer is biocompatible, and the sharp layer is covered to protect the host body.
2. The exposed metal electrode is shielded against corrosion.
3. The feedthroughs that convey the signal into and out of the sealed enclosure are insulated with silicone.

7.7.2   Skin optical loss

The purpose of skin is to shield the inside organs from light. As a result, it becomes difficult to use implanted PV cells to gather energy from ambient light. Light will therefore be diminished as a result of tissue optical loss. Additionally, different skin types will reflect light in various ways [50]. Skin is a complex medium, and temperature, humidity, and radiation cause its optical properties to change dynamically. Additionally, within tissue borders, absorption, scattering, transmission, and reflection are all necessary for light to go through skin. The interfaces between these boundaries are two mediums with differing optical characteristics. The stratum corneum, epidermis, and dermis are the three layers of skin [30,54]. The stratum corneum, also known as the nonliving epidermis (just dead squamous cells), is the outermost layer and ranges in thickness from 10 to 20 μm. With a 20% lipid, 60% protein, and 20% water content, the stratum corneum is highly keratinized. Stratum corneum has a refractive index that ranges from 1.47 to 1.51 [12]. Compared to other tissues, the live epidermis is mostly made up of pigments, the majority of which is melanin. From 1.3% (lightly pigmented) to 43%, melanosomes, a particle used to create melanin, are present (highly pigment) [12]. The amount of pigment varies from person to person, and more pigment might increase melanin absorption. The dermis is known as the main absorber of visible light and is distinguished from the epidermis by a basal lamina. This layer also contains beta-carotene, bilirubin, and blood hemoglobin. The amount of infrared radiation (IR) that is absorbed depends on how much water is present in the dermis [12].

The absorption coefficients of different skin layers are shown in Figure 7.15(a), and the absorptions in the visible spectral range are much higher than those in the IR region. For instance, the absorption coefficient of highly pigmented epidermis is

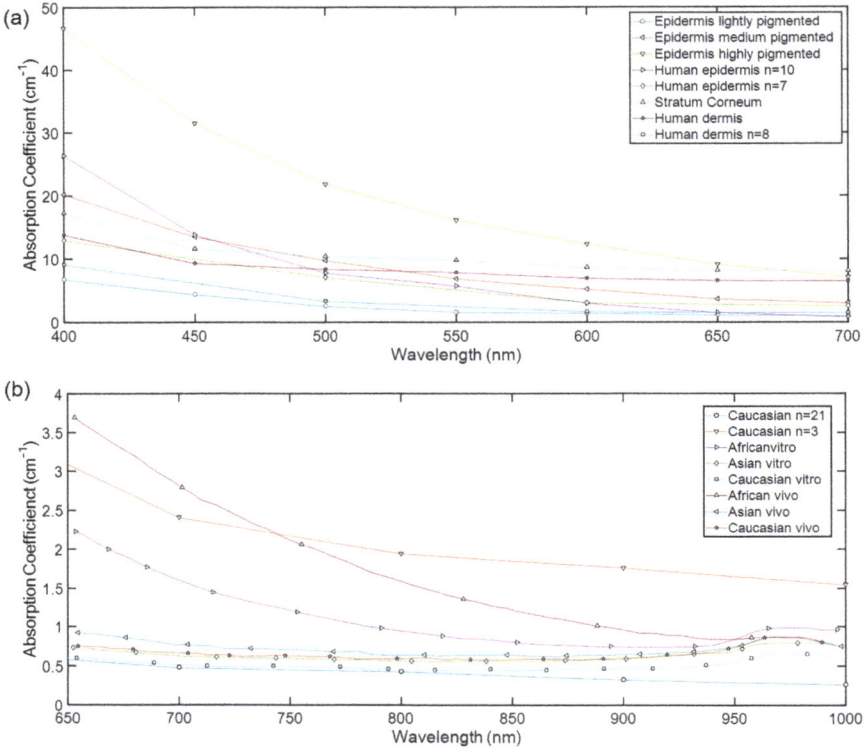

Figure 7.15   Optical coefficients and the human skins. Panel (a) demonstrates how human skin absorbs light at various wavelengths according to its absorption coefficient. In 2011, A. Bashkatov et al. [95] reported on the characteristics of the skin at various depths, from the stratum to the dermis. Panel (b) demonstrates the absorption coefficient for various light spectra. The essay analyses the three ethnic groups, which include Caucasians, Asians, and Africans. The result is from A. Bashkatov et al. [95].

around 50 cm^{-1} at 650 nm while it is 10 cm^{-1} at 700 nm. The skin properties are also different from different ethnic groups, which is shown in Figure 7.15(b).

7.8   Experiment and biomedical experiment in photovoltaic validation

Beyond conventional energy uses, PV technology has demonstrated significant promise and gained new significance in biomedical research, particularly for powering implantable devices. PV cells have a distinct advantage in developing self-sustaining, compact power sources that are necessary for a variety of medical implants because of their capacity to transform light into electrical energy. Thorough testing is necessary to validate PV technology for biomedical applications. This is because PV cells must withstand the complex environment of the human body, which includes fluctuating light intensities, tissue interference, and stringent safety regulations. Evaluation of PV cells' durability, biocompatibility, and energy efficiency in physiologically simulated conditions is another step in this validation process. The advancement of PV-integrated medical devices depends on these experiments, which could result in novel approaches to long-term, low-maintenance power in implants that save lives.

In biomedical research, PV technology – long employed in energy applications – has shown promise, particularly for powering implantable devices. PV cells have a distinct advantage in developing small, self-sufficient power sources that are necessary for a variety of medical implants because of their capacity to transform light into electrical energy. But PV technology needs to be thoroughly tested and validated before it can be used in biological applications. These studies must show that PV cells can operate dependably in the complicated environment of the human body, which includes tight safety regulations, interference from body tissues, and variable light intensities.

Besides offering security in operation, biomedical applications of PV cell validation involve the characterization of every single cell for robustness, biocompatibility, energy efficiency, and under conditions that can realistically simulate physiological situations. Such validation of PV cells will be useful in the study of interactions with body fluids and tissues over a long time and confirm their ability to provide reliable energy for operating the device. The realistic-condition tests will, therefore, form an imperative initial step in assessing the viability of PV cells as reliable long-term power sources for medical implants.

Medical implants could make great progress if PV-integrated devices are validated successfully. This would open up new possibilities for long-lasting, low-maintenance power solutions that support life-saving applications. These developments could improve patients' overall quality of life, reduce the need for intrusive operations to change device batteries, and completely transform the management of chronic diseases. A new generation of energy-autonomous, patient-friendly implants will be made possible by innovations in medical device design spurred by more testing and analysis of PV technology for biomedical implants.

7.8.1   PV validation in biomedical contexts

A systematic strategy to experimentation and validation is required to handle the numerous variables involved as the potential of PV technology in implanted devices keeps expanding. Validation studies need to take into account a PV-powered device's whole life cycle, from implantation to long-term performance under various physiological circumstances. In order to evaluate the PV cells' efficiency in absorbing and converting light over time, experimental settings must precisely imitate the irregularities in light exposure in the body, which changes depending on the location of the device.

Biocompatibility testing is one of the most important phases in PV cell validation for biomedical usage. It is crucial to make sure PV cells don't produce inflammation or unfavorable immune responses because they will be in close proximity to body tissues. In order to assess how PV cells interact with the surrounding tissues and fluids, biocompatibility investigations examine the materials on their surface. Coatings or encapsulating materials are also evaluated to reduce the possibility of adverse immunological reactions. To let light in, these coatings need to be robust, nontoxic, and transparent. Given that any immunological reaction or biofilm formation could impair the cell's ability to function and generate power, long-term studies evaluating cell integration with different tissue types are also essential.

The mechanical forces of the body, including as tissue shifts and motions, must be tolerated by PV cells utilized in implants. Durability testing examines how well PV cells can survive the body's physical stressors over extended durations. In order to verify the PV cells' structural integrity, repetitive motion and pressure variations are simulated. Any possible material deterioration brought on by constant bending, pressure, or chemical exposure may end up needing to be replaced because it will lower the cell's efficiency and hinder device performance. Therefore, to ensure that PV cells can function dependably in the complex, dynamic environment of the body, durability tests include cyclic loading, thermal fluctuation exposure, and testing in fluid environments comparable to those found in the human body.

In low-light conditions, evaluating the energy efficiency and light-to-electricity conversion rate is a significant component of PV validation. Implantable PV cells are exposed to indirect, filtered light that enters subcutaneous or organ-embedded sites, in contrast to conventional PV applications that get direct sunlight. These conditions are replicated in experimental setups by utilizing diffused or filtered light sources to simulate the effects of tissue layers on wavelength and light intensity. To optimize energy conversion, researchers assess the PV cells' performance in these kinds of scenarios and modify the device design as needed. To increase efficiency, novel cell designs and material selections – such as multijunction cells, which catch a wider range of light wavelengths – are also investigated.

Careful integration of power control and energy storage is necessary for implantable PV systems to be feasible. Supercapacitors and microbatteries are examples of storage devices that are evaluated to retain gathered energy and offer a consistent power output to the device because the body's light availability can vary. In order to control this energy, distribute it in accordance with the particular requirements of the device, and preserve energy during dark periods, power management circuits are necessary. To guarantee that PV cells can satisfy the ongoing operational requirements of medical implants, even in the presence of sporadic light exposure, experimental experiments mimic cycles of light availability and energy consumption.

Human considerations must be taken into account in addition to technical validation for successful PV-powered implanted devices. Researchers can better understand how variables like skin thickness, patient movement, and implant placement impact PV cell performance by testing in realistic anatomical models. The placement of the implant – subcutaneously, in the thoracic cavity, or within vascular tissues – may affect the efficacy of PV cells, for example. To obtain information on how each variable affects energy conversion and device operation, experiments may involve putting PV prototypes in cadaveric models or sophisticated tissue-mimicking phantoms that mimic human optical characteristics.

7.8.2 Future directions for PV-integrated biomedical devices

Developing hybrid systems that integrate PV cells with other energy-harvesting methods, increasing cell efficiency, and upgrading the underlying materials are key to the future of PV technology in biomedical applications. There is significant promise for medical applications from research into novel materials like organic PV cells, which can be modified to react better to low light. Additionally, by modifying these organic cells to fit intricate geometries, they might maximize light absorption by closely conforming to skin or organs.

By harnessing various energy sources within the body, hybrid energy systems that integrate PV cells with kinetic or metabolic energy harvesting may also improve reliability. For instance, during exposure times, a PV-kinetic hybrid might harvest mechanical energy from body movements and capture light energy. Patients who are active may benefit most from this kind of technology, which guarantees a steady energy supply even in dimly lit environments. Clinical trials to verify safety and efficacy in human subjects will be the following steps as these inventions develop. Feedback from patients and real-world applications will help refine device designs and tailor them to the unique requirements of various patient populations. Material scientists, engineers, and medical specialists should work closely together to improve PV performance and resolve any relevant issues during clinical application.

With its ability to provide a sustainable and independent energy source for devices that are essential for managing chronic illnesses and enhancing patient quality of life, PV technology has demonstrated previously unheard-of promise in the field of biomedical implants. With extensive testing and validation in physiologically realistic simulations, PV cells are increasingly viable for dependable, long-term power sources inside the body. Continuous improvements in light conversion efficiency, biocompatibility, and material science will make PV-powered implants even more viable. Future developments in PV-integrated devices could pave the way for a new era in individualized, less invasive healthcare by enabling autonomous operation inside the body and lowering reliance on external battery refills. Future implantable devices will increasingly integrate smart energy systems that provide ongoing health monitoring and adaptive therapeutic interventions as research into PV technology advances, ultimately revolutionizing patient care and health outcomes.

References

[1]   S. Abdellatif, K. Kirah, R. Ghannam, A.S.G. Khalil, and W. Anis. Enhancing the absorption capabilities of thin-film solar cells using sandwiched light trapping structures. *Applied Optics*, 54(17):5534–5541, 2015.

[2]   H. Abdulsalam, G. Babaji, and H.T. Abba. The effect of temperature and active layer thickness on the performance of CH3NH3PbI3 perovskite

solar cell: A numerical simulation approach. *Journal for Foundations and Applications of Physics*, 5(2):141–151, 2018.

[3]   S.-H. Ahn, J. Jeong, and S.J. Kim. Emerging encapsulation technologies for long-term reliability of microfabricated implantable devices. *Micromachines*, 10(8):508, 2019.

[4]   A. Ahnood, K.E. Fox, N.V. Apollo, *et al.* Diamond encapsulated photovoltaics for transdermal power delivery. *Biosensors and Bioelectronics*, 77:589–597, 2016.

[5]   P. Altermatt. Models for numerical device simulations of crystalline silicon solar cells – A review. *Journal of Computational Electronics*, 10:314–330, 2011.

[6]   P. Altermatt. *Numerical Simulation of Crystalline Silicon Solar Cells*, chapter 3.8, pages 150–159. John Wiley & Sons, Ltd, 2016.

[7]   P.P. Altermatt, J.O. Schumacher, A. Cuevas, *et al.* Numerical modeling of highly doped Si: P emitters based on Fermi–Dirac statistics and self-consistent material parameters. *Journal of Applied Physics*, 92(6):3187–3197, 2002.

[8]   Y. Arima and M. Ehara. On-chip solar battery structure for CMOS LSI. *IEICE Electronics Express*, 3(13):287–291, 2006.

[9]   D.E. Aspnes and A.A. Studna. Dielectric functions and optical parameters of Si, Ge, GaP, GaAs, GaSb, InP, InAs, and InSb from 1.5 to 6.0 ev. *Physical Review B*, 27:985–1009, 1983.

[10]   H.A. Atwater. Seeing the light in energy use. *Nanophotonics*, 10(1): 115–116, 2021.

[11]   S. Ayazian, V.A. Akhavan, E. Soenen, and A. Hassibi. A photovoltaic-driven and energy-autonomous CMOS implantable sensor. *IEEE Transactions on Biomedical Circuits and Systems*, 6(4):336–343, 2012.

[12]   A.N. Bashkatov, E.A. Genina, and V.V. Tuchin. Optical properties of skin, subcutaneous, and muscle tissues: A review. *Journal of Innovative Optical Health Sciences*, 4(01):9–38, 2011.

[13]   P.A. Basore. Numerical modeling of textured silicon solar cells using PC-1D. *IEEE Transactions on Electron Devices*, 37(2):337–343, 1990.

[14]   P.A. Basore. Efficient computation of multidimensional Lambertian optical absorption. *IEEE Journal of Photovoltaics*, 9(1):106–111, 2018.

[15]   G. Bo, L. Ren, X. Xu, Y. Du, and S. Dou. Recent progress on liquid metals and their applications. *Advances in Physics: X*, 3(1):1446359, 2018.

[16]   M. Burgelman, P. Nollet, and S. Degrave. Modelling polycrystalline semiconductor solar cells. *Thin Solid Films*, 361:527–532, 2000.

[17]   J.-F. Chen, C.-L. Chun, and Y.-J. Hung. Mirror-assisted interdigitated back-contact CMOS photovoltaic devices for powering subcutaneous implantable devices. In *2015 International Symposium on Next-Generation Electronics (ISNE)*, pages 1–4. IEEE, 2015.

[18]   Z. Chen, M.-K. Law, P.-I. Mak, and R.P. Martins. A single-chip solar energy harvesting IC using integrated photodiodes for biomedical implant

applications. *IEEE Transactions on Biomedical Circuits and Systems*, 11(1):44–53, 2016.

[19] J.H. Choi, Y. Mao, and J.P. Chang. Development of Hafnium based high-k materials – A review. *Materials Science and Engineering: R: Reports*, 72(6):97–136, 2011.

[20] P.E. Ciddor. Refractive index of air: New equations for the visible and near infrared. *Applied Optics*, 35(9):1566–1573, 1996.

[21] R. Couderc, M. Amara, and M. Lemiti. Reassessment of the intrinsic carrier density temperature dependence in crystalline silicon. *Journal of Applied Physics*, 115(9):093705, 2014.

[22] A.V. Dan T.-I. and B.G. Min. Design of the solar cell system for recharging the external battery of the totally-implantable artificial heart. *The International Journal of Artificial Organs*, 22(12):823–826, 1999.

[23] T. Das, S. Prasad, S. Dam, and P. Mandal. A pseudo cross-coupled switch-capacitor based DC–DC boost converter for high efficiency and high power density. *IEEE Transactions on Power Electronics*, 29(11):5961–5974, 2014.

[24] I. Dion, C. Baquey, and J.R. Monties. Diamond: The biomaterial of the 21st century? *The International Journal of Artificial Organs*, 16(9):623–627, 1993.

[25] L. El Chaar, L.A. lamont, and N. El Zein. Review of photovoltaic technologies. *Renewable and Sustainable Energy Reviews*, 15(5):2165–2175, 2011.

[26] L. Eldada, F. Adurodija, B. Sang, *et al.* Development of hybrid copper indium gallium selenide photovoltaic devices by the FASST® printing process. In *Proceedings of the European Photovoltaic Solar Energy Conference*, volume 23, page 2142, 2008.

[27] J.G. Fossum. Computer-aided numerical analysis of silicon solar cells. *Solid-State Electronics*, 19(4):269–277, 1976.

[28] J.G. Fossum. Physical operation of back-surface-field silicon solar cells. *IEEE Transactions on Electron Devices*, 24(4):322–325, 1977.

[29] A. Froitzheim, R. Stangl, L. Elstner, M. Kriegel, and W. Fuhs. AFORS-HET: A computer-program for the simulation of heterojunction solar cells to be distributed for public use. In *3rd World Conference on Photovoltaic Energy Conversion*, volume 1, pages 279–282, 2003.

[30] M. Geerligs. Skin layer mechanics. *Eindhoven: TU Eindhoven*, 2010.

[31] M. Ghannam and Y. Abdulraheem. Electro-physical interpretation of the degradation of the fill factor of silicon heterojunction solar cells due to incomplete hole collection at the a-Si: H/c-Si thermionic emission barrier. *Applied Sciences*, 8(10):1846, 2018.

[32] R. Ghannam, P.V. Klaine, and M. Imran. Artificial intelligence for photovoltaic systems. In *Solar Photovoltaic Power Plants*, pages 121–142. Springer, 2019.

[33] G. Ghosh. Dispersion-equation coefficients for the refractive index and birefringence of calcite and quartz crystals. *Optics Communications*, 163(1–3):95–102, 1999.

[34]   G. Gibbons and S.M. Sze. Avalanche breakdown voltages of abrupt and lin-
        early graded p–n junctions in Ge, Si, GaAs, and GaP (avalanche breakdown
        voltages of abrupt and linearly graded p–n junction in Ge, Si, GaAs, and
        GaP). *Applied Physics Letters*, 8:111–113, 1966.

[35]   S. Gong and W. Cheng. Toward soft skin-like wearable and implantable
        energy devices. *Advanced Energy Materials*, 7(23):1700648, 2017.

[36]   J.L. Gray. ADEPT: A general purpose numerical device simulator for mod-
        eling solar cells in one-, two-, and three-dimensions. In *The Conference
        Record of the Twenty-Second IEEE Photovoltaic Specialists Conference-
        1991*, pages 436–438. IEEE, 1991.

[37]   M.A. Green, Y. Hishikawa, E.D. Dunlop, D.H. Levi, J. Hohl-Ebinger, and
        A.W.Y. Ho-Baillie. Solar cell efficiency tables (Version 52). *Progress in
        Photovoltaics: Research and Applications*, 26(7):427–436, 2018.

[38]   A.S. Grove. *Physics and Technology of Semiconductor Devices*. Wiley, 1967.

[39]   A. Haeberlin, A. Zurbuchen, J. Schaerer, *et al.* Successful pacing using
        a batteryless sunlight-powered pacemaker. *Europace*, 16(10):1534–1539,
        2014.

[40]   A. Haeberlin, A. Zurbuchen, S. Walpen, *et al.* The first batteryless, solar-
        powered cardiac pacemaker. *Heart Rhythm*, 12(6):1317–1323, 2015.

[41]   M.A. Hannan, S. Mutashar, S.A. Samad, and A. Hussain. Energy harvesting
        for the implantable biomedical devices: Issues and challenges. *Biomedical
        Engineering Online*, 13(1):79, 2014.

[42]   G. Hashmi, A.R. Akand, M. Hoq, and H. Rahman. Study of the enhance-
        ment of the efficiency of the monocrystalline silicon solar cell by optimiz-
        ing effective parameters using PC1D simulation. *Silicon*, 10(4):1653–1660,
        2018.

[43]   G. Hong and G. Han. Design optimization of photovoltaic cell stacking
        in a triple-well CMOS process. *IEEE Transactions on Electron Devices*,
        67(6):2381–2385, 2020.

[44]   K.O. Htet, R. Ghannam, Q. H. Abbasi, and H. Heidari. Power management
        using photovoltaic cells for implantable devices. *IEEE Access*, 6:42156–
        42164, 2018. DOI: 10.1109/ACCESS.2018.2860793

[45]   Z. Hua and H. Lee. A reconfigurable dual-output switched-capacitor DC–
        DC regulator with sub-harmonic adaptive-on-time control for low-power
        applications. *IEEE Journal of Solid-State Circuits*, 50(3):724–736, 2015.

[46]   C.-H. Huang and W.-J. Chuang. Dependence of performance parameters of
        CdTe solar cells on semiconductor properties studied by using SCAPS-1D.
        Vacuum, 118:32–37, 2015.

[47]   J. Humlíček and K. Vojtěchovský. Infrared optical constants of n-type
        silicon. *Czechoslovak Journal of Physics B*, 38(9):1033–1049, 1988.

[48]   Y.-Jr Hung, M.-S. Cai, J.-F. *et al.* High-voltage backside-illuminated CMOS
        photovoltaic module for powering implantable temperature sensors. *IEEE
        Journal of Photovoltaics*, 8(1):342–347, 2017.

[49]   Y.-Jr Hung, T.-Y. Chuang, C.-L. Chun, M.-S. Cai, H.-W. Su, and S.-L.
        Lee. CMOS-enabled interdigitated back-contact solar cells for biomedical

applications. *IEEE Transactions on Electron Devices*, 61(12):4019–4024, 2014.

[50] S.L. Jacques. Optical properties of biological tissues: A review. *Physics in Medicine & Biology*, 58(11):R37, 2013.

[51] Y. Jiang, M.-K. Law, Z. Chen, P.-I. Mak, and R.P. Martins. Algebraic series-parallel-based switched-capacitor DC–DC boost converter with wide input voltage range and enhanced power density. *IEEE Journal of Solid-State Circuits*, 54(11):3118–3134, 2019.

[52] M.D. Kelzenberg, M.C. Putnam, D.B. Turner-Evans, N.S. Lewis, and H.A. Atwater. Predicted efficiency of Si wire array solar cells. In *2009 34th IEEE Photovoltaic Specialists Conference (PVSC)*, pages 001948–001953, 2009.

[53] B. Ketola, K.R. McIntosh, A. Norris, *et al.* Silicones for photovoltaic encapsulation. In *23rd European Photovoltaic Solar Energy Conference*, pages 1–5, 2008.

[54] Pa.A.J. Kolarsick, M.A. Kolarsick, and C. Goodwin. Anatomy and physiology of the skin. *Journal of the Dermatology Nurses' Association*, 3(4):203–213, 2011.

[55] T. Laube, C. Brockmann, R. Buß, *et al.* Optical energy transfer for intraocular microsystems studied in rabbits. *Graefe's Archive for Clinical and Experimental Ophthalmology*, 242(8):661–667, 2004.

[56] M.K. Law and A. Bermak. High-voltage generation with stacked photodiodes in standard CMOS process. *IEEE Electron Device Letters*, 31(12):1425–1427, 2010.

[57] M. Levinshtein, S. Rumyantsev, and M Shur. *Handbook Series on Semiconductor Parameters*, volume 1. Singapore: World Scientific, 1996.

[58] J. Li, C.-H. Wen, S. Gauza, R. Lu, and S.-T. Wu. Refractive indices of liquid crystals for display applications. *Journal of Display Technology*, 1(1):51, 2005.

[59] X. Li, N.P. Hylton, V. Giannini, K.-H. Lee, N.J. Ekins-Daukes, and S.A. Maier. Multi-dimensional modeling of solar cells with electromagnetic and carrier transport calculations. *Progress in Photovoltaics: Research and Applications*, 21(1):109–120, 2013.

[60] X. Liu, L. Huang, K. Ravichandran, and E. Sánchez-Sinencio. A highly efficient reconfigurable charge pump energy harvester with wide harvesting range and two-dimensional MPPT for internet of things. *IEEE Journal of Solid-State Circuits*, 51(5):1302–1312, 2016.

[61] L. Lu, Z. Yang, K. Meacham, *et al.* Biodegradable monocrystalline silicon photovoltaic microcells as power supplies for transient biomedical implants. *Advanced Energy Materials*, 8(16):1703035, 2018.

[62] Y. Lu, J. Jiang, and W.-H. Ki. A multiphase switched-capacitor DC–DC converter ring with fast transient response and small ripple. *IEEE Journal of Solid-State Circuits*, 52(2):579–591, 2016.

[63] T.K. Manna and S.M. Mahajan. Nanotechnology in the development of photovoltaic cells. In *2007 International Conference on Clean Electrical Power*, pages 379–386. IEEE, 2007.

[64]   B. Mohammadi and J. Rodrigues. Ultra low energy and area efficient charge pump with automatic clock controller in 65 nm CMOS. In *2015 IEEE Asian Solid-State Circuits Conference (A-SSCC)*, pages 1–4. IEEE, 2015.

[65]   S. Mondal and R. Paily. Efficient solar power management system for self-powered IoT node. *IEEE Transactions on Circuits and Systems I: Regular Papers*, 64(9):2359–2369, 2017.

[66]   S. Mondal and R. Paily. On-chip photovoltaic power harvesting system with low-overhead adaptive MPPT for IoT nodes. *IEEE Internet of Things Journal*, 4(5):1624–1633, 2017.

[67]   S. Mondal and R.P. Paily. An efficient on chip power management architecture for solar energy harvesting systems. In *2016 29th International Conference on VLSI Design and 2016 15th International Conference on Embedded Systems (VLSID)*, pages 224–229. IEEE, 2016.

[68]   E. Moon, D. Blaauw, and J.D. Phillips. Infrared energy harvesting in millimeter-scale GaAs photovoltaics. *IEEE Transactions on Electron Devices*, 64(11):4554–4560, 2017.

[69]   E. Moon, D. Blaauw, and J.D. Phillips. Subcutaneous photovoltaic infrared energy harvesting for bio-implantable devices. *IEEE Transactions on Electron Devices*, 64(5):2432–2437, 2017.

[70]   M. Mostefaoui, H. Mazari, S. Khelifi, A. Bouraiou, and R. Dabou. Simulation of high efficiency CIGS solar cells with SCAPS-1D software. *Energy Procedia*, 74:736–744, 2015.

[71]   J. Nelson. *The physics of solar cells*. Imperial College Press, 2003.

[72]   H.T. Nguyen, S.C. Baker-Finch, and D. Macdonald. Temperature dependence of the radiative recombination coefficient in crystalline silicon from spectral photoluminescence. *Applied Physics Letters*, 104(11):112105, 2014.

[73]   T.P. Otanicar, P.E. Phelan, and J.S. Golden. Optical properties of liquids for direct absorption solar thermal energy systems. *Solar Energy*, 83(7): 969–977, 2009.

[74]   H. Peng, N. Tang, Y. Yang, and D. Heo. CMOS startup charge pump with body bias and backward control for energy harvesting step-up converters. *IEEE Transactions on Circuits and Systems I: Regular Papers*, 61(6): 1618–1628, 2014.

[75]   H.R. Phillip and E.A. Taft. Kramers-Kronig analysis of reflectance data for diamond. *Physical Review*, 136(5A):A1445, 1964.

[76]   E. Płaczek-Popko. Top PV market solar cells 2016. *Opto-Electronics Review*, 25(2):55–64, 2017.

[77]   S. Popova, T. Tolstykh, and V. Vorobev. Optical characteristics of amorphous quartz in the 1400-200 cm- 1 region. *Opt. Spectrosc*, 33:444–445, 1972.

[78]   M.R. Querry. Optical constants. Technical Report, MISSOURI UNIV-KANSAS CITY, 1985.

[79]   J.R. Ray, C.J. Panchal, M.S. Desai, and U.B. Trivedi. Simulation of CIGS thin film solar cells using AMPS-1D. *Journal of Nano- and Electronic Physics*, 3(1):747–754, 2011.

[80] M. Riaz, A.C. Kadhim, S.K. Earles, and A. Azzahrani. Variation in efficiency with change in band gap and thickness in thin film amorphous silicon tandem heterojunction solar cells with AFROS-HET. *Optics Express*, 26(14): A626–A635, 2018.

[81] L.V. Rodríguez-de Marcos, J.I. Larruquert, J.A. Méndez, and J.A. Aznárez. Self-consistent optical constants of SiO2 and Ta2O5 films. *Optical Materials Express*, 6(11):3622–3637, 2016.

[82] A. Rohatgi and P. Rai-Choudhury. Design, fabrication, and analysis of 17-18-percent efficient surface-passivated silicon solar cells. *IEEE Transactions on Electron Devices*, 31(5):596–601, 1984.

[83] M.W. Rowell, M.A. Topinka, M.D. McGehee, *et al.* Organic solar cells with carbon nanotube network electrodes. *Applied Physics Letters*, 88(23):233506, 2006.

[84] B. Rumberg, D.W. Graham, and M.M. Navidi. A regulated charge pump for tunneling floating-gate transistors. *IEEE Transactions on Circuits and Systems I: Regular Papers*, 64(3):516–527, 2017.

[85] K. Shen and M.M. Maharbiz. Ceramic packages for acoustically coupled neural implants. In *2019 9th International IEEE/EMBS Conference on Neural Engineering (NER)*, pages 847–850. IEEE, 2019.

[86] X. Sheng, S. Wang, and L. Yin. Flexible, stretchable, and biodegradable thin-film silicon photovoltaics. In *Advances in Silicon Solar Cells*, pages 161–175. Springer, 2018.

[87] G. Simone, D.D.C. Rasi, X. de Vries, *et al.* Near-infrared tandem organic photodiodes for future application in artificial retinal implants. *Advanced Materials*, 30(51):1804678, 2018.

[88] S.B. Sohid and A. Kabalan. Numerical analysis of ZnTe based solar cell with Sb2Te3 back surface field layer using SCAPS-1D. In *2018 IEEE 7th World Conference on Photovoltaic Energy Conversion (WCPEC) (A Joint Conference of 45th IEEE PVSC, 28th PVSEC 34th EU PVSEC)*, pages 1852–1857, June 2018.

[89] K. Song, J.H. Han, T. Lim, *et al.* Subdermal flexible solar cell arrays for powering medical electronic implants. *Advanced healthcare materials*, 5(13):1572–1580, 2016.

[90] K. Song, J.H. Han, H.C. Yang, K.I. Nam, and J. Lee. Generation of electrical power under human skin by subdermal solar cell arrays for implantable bioelectronic devices. *Biosensors and Bioelectronics*, 92:364–371, 2017.

[91] S. Strehlke, S. Bastide, and C. Lévy-Clément. Optimization of porous silicon reflectance for silicon photovoltaic cells. *Solar Energy Materials and Solar Cells*, 58(4):399–409, 1999.

[92] N.A. Tegegne and F.G. Hone. Solar cell technology: Challenges and progress. In *Electrode Materials for Energy Storage and Conversion*, pages 437–471. CRC Press, 2021.

[93] T. Tokuda, T. Ishizu, W. Nattakarn, *et al.* 1 mm^3-sized optical neural stimulator based on CMOS integrated photovoltaic power receiver. *AIP Advances*, 8(4):045018, 2018.

[94]   W. Van Roosbroeck. Theory of the flow of electrons and holes in germanium and other semiconductors. *The Bell System Technical Journal*, 29(4):560–607, 1950.

[95]   M.R. Vogt, H. Holst, H. Schulte-Huxel, *et al.* Optical constants of UV transparent EVA and the impact on the PV module output power under realistic irradiation. *Energy Procedia*, 92:523–530, 2016.

[96]   G. Vuye, S. Fisson, V. Nguyen Van, Y. Wang, J. Rivory, and F. Abeles. Temperature dependence of the dielectric function of silicon using in situ spectroscopic ellipsometry. *Thin Solid Films*, 233(1–2):166–170, 1993.

[97]   F. Wang, H. Yu, J. Li, *et al.* Design guideline of high efficiency crystalline Si thin film solar cell with nanohole array textured surface. *Journal of Applied Physics*, 109(8):084306, 2011.

[98]   M.J. Weber. *Handbook of Optical Materials*, volume 19. CRC press, 2002.

[99]   S.M. Wong, H.Y. Yu, Y. Li, *et al.* Boosting short-circuit current with rationally designed periodic Si nanopillar surface texturing for solar cells. *IEEE Transactions on Electron Devices*, 58(9):3224–3229, 2011.

[100] D.L. Wood, K. Nassau, T.Y. Kometani, and D.L. Nash. Optical properties of cubic hafnia stabilized with yttria. *Applied Optics*, 29(4):604–607, 1990.

[101] T. Wu, J.M. Redouté, and M.R. Yuce. A wireless implantable sensor design with subcutaneous energy harvesting for long-term IoT healthcare applications. *IEEE Access*, 6:35801–35808, 2018.

[102] X. Wu, Y. Shi, S. Jeloka, *et al.* A 20-pw discontinuous switched-capacitor energy harvester for smart sensor applications. *IEEE Journal of Solid-State Circuits*, 52(4):972–984, 2017.

[103] C. Yan and P.S. Lee. Stretchable energy storage and conversion devices. *Small*, 10(17):3443–3460, 2014.

[104] S. Yoon, S. Carreon-Bautista, and E. Sánchez-Sinencio. An area efficient thermal energy harvester with reconfigurable capacitor charge pump for IoT applications. *IEEE Transactions on Circuits and Systems II: Express Briefs*, 65(12):1974–1978, 2018.

[105] J. Zhao, R. Ghannam, M.K. Law, M.A. Imran, and H. Heidari. Photovoltaic power harvesting technologies in biomedical implantable devices considering the optimal location. *IEEE Journal of Electromagnetics, RF and Microwaves in Medicine and Biology*, 4(2):148–155, 2020.

[106] J. Zhao, K.O. Htet, R. Ghannam, M. Imran, and H. Heidari. Modelling of implantable photovoltaic cells based on human skin types. In *2019 15th Conference on Ph. D Research in Microelectronics and Electronics (PRIME)*, pages 253–256. IEEE, 2019.

Chapter 8
Hybrid energy harvesting

8.1 Piezoelectric–electromagnetic (PE–EM) hybrid systems

Piezoelectric (PE) and electromagnetic (EM) hybrid energy harvesters provide a possible solution to increase the power requirements of implantable medical devices using body environmental vibration energy. Hybrid systems can provide higher energy conversion efficiency, increased power density, and flexibility to a broader range of vibration frequencies within the human body due to the combined benefits of both PE and EM mechanisms. This combination is particularly useful in situations where energy availability is intermittent or comes from low-frequency physiological functions such as breathing, heartbeat, and muscular action [3].

Hybrid energy harvesters may utilize PE and EM transducers on the same structure, usually a cantilever beam or a bimorph system, which maximizes the output power over a greater variety of vibration conditions. Under bending conditions, the PE layer generates charge through its associated strain, and further extra power is delivered because of the relative motion between the magnet and the coil via EM induction. This dual harvesting approach allows the system to detect both high-frequency, low-amplitude vibrations and low-frequency, larger displacements that are common in implantable environments. The schematic of two-pole and four-pole PE–energy harvester (EH) energy harvesting system are shown in Figure 8.1.

Advanced design methodologies can further improve the performance of PE–EM hybrid energy harvesters for implantable applications. The use of a trapezoidal or rectangular cantilever construction with an optimum tip mass (magnet for EM coupling) enhances the energy harvesting capability. Resonance at desirable physiological frequencies can be achieved with the device through precise tuning of beam size and material properties. Bimorph and multilayer configurations, having two active PE layers on opposite sides of the substrate, provide higher power compared with unimorph systems. The EM coils in hybrid systems are strategically placed in order to reduce coupling loss and enhance induction efficiency. Generally, hybrid systems separate PE and EM transducers, with a view to minimize interference and consequently maximize the energy extracted. Research has shown that the use of parallel-connected PE–EM harvesters produces three times more power output compared to serial connections, and thus it would be more ideal for IMDs. The application of high-efficiency PE materials such as lead zirconate titanate (PZT) and sophisticated

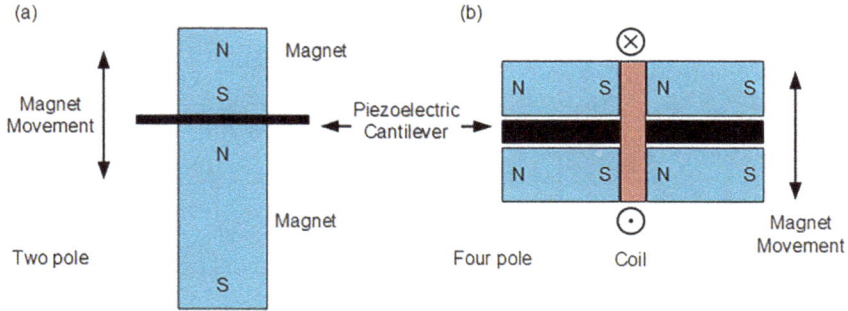

Figure 8.1   (a) Two-pole PE–EH energy harvesting system [2,3]. (b) Four-pole PE–EH energy harvesting system [1,3].

EM components in hybrid systems leads to improved conversion efficiencies. Biocompatible encapsulation materials ensure not only safe functioning within the body but also protection from biological fluids [3].

The experimental results of various investigations demonstrate the possibility of implantable applications of PE–EM hybrid harvesters. For instance, some tests conducted with prototypes having PZT layers and EM transducers showed that in the case of low-frequency vibrations of 50 Hz and with accelerations as low as 0.4g, the total power output is 10.7 mW. With optimized magnet arrangements and cantilever designs, hybrid harvesters produced much higher power densities than PE or EM harvesters individually. Furthermore, tests demonstrate that hybrid systems with four-pole magnet arrangements can reach up to 3.5 mW at 1g excitation levels, considerably extending standard energy harvesting techniques [3].

Linear PE–EM systems are designed to operate within predictable vibration patterns; the production of energy is thus mainly determined by vibration frequency, damping ratios, and mechanical and electrical resonance alignment. In 2013, Ping Li *et al.* conducted research into the performance of linear hybrid energy harvester (HEH) under white-noise excitation. Their results indicated that in PE–EM system, vibration frequency and coupling coefficients have a great influence on power generation because they determine bandwidth and optimum power. Unlike conventional systems, PE–EM systems offer the advantage of impedance matching for optimal performance over a wide range of vibration frequencies. In the linear PE–EM system, no fixed magnets are utilized, making the design simpler yet still strong in performance [6].

The addition of frequency tweaking mechanisms has improved the versatility of PE energy-harvesting devices. Wischke *et al.* investigated the frequency tuning strategy in PE–EH systems using voltage-controlled stiffness variations of the PE layer, and it directly affects the resonance frequency of the system [10]. The input excitation synchronized with the resonance of the generator significantly obtained the broadband operation in PE–EH systems. This tunability is particularly helpful for systems that experience fluctuating mechanical inputs because static designs have a lot of inconsistency in power. For example, experiments revealed that when PE beams are extended beyond a length of 10 mm, the maximum adjustable frequency saturation occurs at approximately 50–60 Hz, improving bandwidth operation. In parallel configurations, the PE component can generate 200 μW while the EM transducer can supply up to 50 μW at frequencies of approximately 56 Hz [11].

To widen the bandwidth and enhance the efficiency, nonlinear PE–EH systems have been presented. Nonlinear mechanisms introduce dynamic behaviors that broaden the operational frequency range of the system and thus allow energy harvesting over a wider range. Li *et al.* highlighted key advantages of the nonlinear system, such as higher bandwidth at reduced resonance frequencies and increased power output as acceleration increases [4]. Shan *et al.*, among others, have proposed single-stable PE–EH systems that have opposing facing magnets, which create the nonlinear response [9]. This configuration yielded maximum values of 11.4 and 21.6 mW at 8.37 and 14.83 Hz, respectively, while being able to operate within an effective frequency range of between 7 and 17 Hz.

Xu *et al.* improved the nonlinear PE–EH principle by setting aligned magnets close to the cantilever tip; hence, a larger operational frequency band was obtained. Its prototype generated a higher power output of 5.66 mW at 1*g* acceleration, which was 247% higher than output at lower excitation levels [13]. Moreover, the bandwidth was increased by 83.3%, showing that nonlinear PE–EH systems can provide larger energy output while maintaining stability for erratic mechanical input.

The researchers compared the performance of PE–EH systems with that of standalone PE or EM systems. Mahmoudi *et al.* proved that isolated PE and EM components in hybrid systems can independently contribute to energy harvesting, resulting in much higher power densities [8]. This study achieved a power density gain of up to 60% and a bandwidth increase of 29% in the frequency range of 155–220 Hz by combining bimorph PE layers with PZT materials. The combination of PZT-based PE transducers with EM coils is especially effective in addressing low-frequency mechanical vibrations and producing consistent energy output within changing environmental conditions.

Ping Li *et al.* further developed these works incorporating advanced nonlinear elements and conducting extensive research on load optimization, input frequency matching, and dynamic amplification [7]. The hybrid PE–EH prototype achieved 3.6 mW of output power at 110 Hz, with a half-power bandwidth of 107.5–112.5 Hz. Greater bandwidth, combined with higher power generation, shows that the PE–EH system is able to outperform standard energy harvesters in complicated vibration scenarios.

The versatility and high efficiency of PE–EH systems point toward their suitability for use in implantable energy harvesting systems that involve a consistent generation of power. To date, PE–EH systems have been coupled into medical devices such as cardiac pacemakers and neurostimulators, and even in low-frequency motion by extracting energy. Since energy conversion using the combined technique will allow for a steady amount of energy output during irregularities in motion or low amplitude, it will further minimize the need for battery replacement, improving the lifetime of devices [7,12].

Frequency tuning, nonlinear topologies, and optimal material selection are only a few of the continuous developments in PE–EH system design that show great potential for next-generation energy harvesting solutions. PE–EH systems are expected to overcome the present limitations and deliver high-efficiency sustainable energy solutions for implantable medical devices and other upcoming applications [7,8,11,12].

8.2   Other hybrid energy harvesting systems

Gambier *et al.* presented a new hybrid energy harvester combining solar transducers, PE generators, and thermoelectric generators in a multisource energy harvester system [3,5]. This multiform approach took advantage of the different positive contributions of various energy-to-energy conversion mechanisms to further improve the global power generation efficiency of a harvester.

The hybrid system here proposed is realized using flexible PV cells PowerFilm, Inc., piezoceramic generators QuickPack QP10n, Mide TC, and thin-film batteries. These elements were fabricated onto metallic substrates that were coated with Kapton layers for exceptional mechanical robustness and electrical isolation. Additional embedded flexible copper electrodes increased the operating reliability of the system, as illustrated in Figure 8.2 [5]. The single-layer solar transducer of the system is $93 \times 25 \times 0.178$ mm^3 while the cantilever-based PE generator is $93 \times 25 \times 1.5$ mm^3. Each energy harvesting mechanism was independently tested along with combined operation at light intensities of 124, 223, 311, and 437 W/m^2. The system was able to give a maximum output power at the highest irradiance condition of 437 W/m^2 with a peak value at 30 mW when the irradiance level dropped to a value of 223 W/m^2, the average power became 12.5 mW [5]. Frequency response testing further characterizes the dynamic behavior of the PE transducer. For this, an excitation frequency sweep, starting from 0–500 Hz, shows a resonance peak at 56.4 Hz. At this frequency, the PE transducer provided power outputs of 0.4 and 0.49 mW for base excitations of $0.1g$ and $0.5g$, respectively [5]. Hehr *et al.* also presented a hybrid energy harvester that harvested both vibrational and radio frequency (RF) energy using a single solenoid coil [5]. The electromagnetic harvester (EMH) is made up of a solenoid coil that is 50.8 mm in diameter and 95.3 mm in length, as well as a neodymium magnet

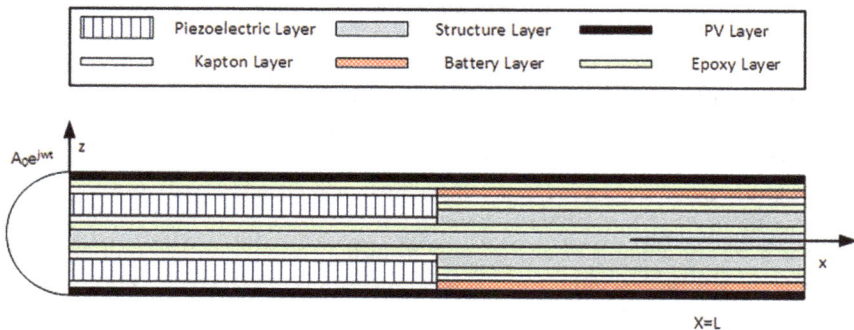

Figure 8.2   Multilayer hybrid energy harvester [3,5]

that is 38.1 mm in diameter and 12.5 mm long. The device reached an initial resonance frequency of 12.5 Hz, which was experimentally validated using acceleration amplitudes of 3.5*g* and 5*g* [5].

A coil diameter of 40.5 mm was chosen to maximize RF energy collection capability, corresponding to a center frequency of 2.45 GHz, excellent for the industrial, scientific, and medical RF bands. EM resonance testing showed that performance was verified at 2.46 GHz. For the highest voltage output, a helical antenna with a 39-mm diameter grounding copper ring was used [5].

The proposed multi-source hybrid energy harvester derives significant efficiency and flexibility benefits from its PV, PE, thermoelectric, and RF energy collecting capability. Such systems have a huge potential for microelectronics, wireless sensor networks, and other low-power applications that demand a long-term, self-sustained energy source.

References

[1]   M.F. Ab Rahman, S.L. Kok, N.M. Ali, R.A. Hamzah, and K.A.A. Aziz. Hybrid vibration energy harvester based on piezoelectric and electromagnetic transduction mechanism. In *2013 IEEE Conference on Clean Energy and Technology (CEAT)*, pages 243–247. Piscataway, NJ: IEEE, 2013.

[2]   M.F. Ab Rahman, S.L. Kok, E. Ruslan, A.H. Dahalan, and S. Salam. Comparison study between four poles and two poles magnets structure in the hybrid vibration energy harvester. In *2013 IEEE Student Conference on Research and Development*, pages 227–231. Piscataway, NJ: IEEE, 2013.

[3]   N. Bizon, N.M. Tabatabaei, F. Blaabjerg, and E. Kurt. Energy harvesting and energy efficiency. *Technology, Methods, and Applications*, 37, 2017.

[4]   D. Castagnetti. A Belleville-spring-based electromagnetic energy harvester. *Smart Materials and Structures*, 24(9):094009, 2015.

[5]   P. Gambier, S.R. Anton, N. Kong, A. Erturk, and D.J. Inman. Piezoelectric, solar and thermal energy harvesting for hybrid low-power generator systems with thin-film batteries. *Measurement Science and Technology*, 23(1):015101, 2011.

[6]   P. Li, S. Gao, H. Cai, and L. Wu. Theoretical analysis and experimental study for nonlinear hybrid piezoelectric and electromagnetic energy harvester. *Microsystem Technologies*, 22:727–739, 2016.

[7]   P. Li, S. Gao, H. Cai, and L. Wu. Theoretical analysis and experimental study for nonlinear hybrid piezoelectric and electromagnetic energy harvester. *Microsystem Technologies*, 22:727–739, 2016.

[8]   S. Mahmoudi, N. Kacem, and N. Bouhaddi. Enhancement of the performance of a hybrid nonlinear vibration energy harvester based on piezoelectric and electromagnetic transductions. *Smart Materials and Structures*, 23(7):075024, 2014.

[9]   X.-B. Shan, S.-W. Guan, Z.-S. Liu, Z.-L. Xu, and T. Xie. A new energy harvester using a piezoelectric and suspension electromagnetic mechanism. *Journal of Zhejiang University SCIENCE A*, 14(12):890–897, 2013.

[10]   M. Wischke, M. Masur, F. Goldschmidtboeing, and P. Woias. Electromagnetic vibration harvester with piezoelectrically tunable resonance frequency. *Journal of Micromechanics and Microengineering*, 20(3):035025, 2010.
[11]   M. Wischke, M. Masur, F. Goldschmidtboeing, and P. Woias. Piezoelectrically tunable electromagnetic vibration harvester. In *2010 IEEE 23rd International Conference on Micro Electro Mechanical Systems (MEMS)*, pages 1199–1202. Piscataway, NJ: IEEE, 2010.
[12]   H. Xia, R. Chen, and L. Ren. Analysis of piezoelectric–electromagnetic hybrid vibration energy harvester under different electrical boundary conditions. *Sensors and Actuators A: Physical*, 234:87–98, 2015.
[13]   Z.L. Xu, X.B. Shan, R.J. Song, and T. Xie. Electromechanical modeling and experimental verification of nonlinear hybrid vibration energy harvester. In *2014 Joint IEEE International Symposium on the Applications of Ferroelectric, International Workshop on Acoustic Transduction Materials and Devices & Workshop on Piezoresponse Force Microscopy*, pages 1–4. Piscataway, NJ: IEEE, 2014.

Chapter 9
Implantable applications with energy harvesting technology

In recent years, implantable medical devices (IMDs) have expanded to encompass a wide variety of devices, such as neural stimulators, sensory prosthetics, drug delivery, and synthetic organs, with each class of devices exhibiting distinct energy requirements. In this chapter, we first review various classes of cardiac pacemakers and biosensors that typically have energy requirements ranging from microwatts to several milliwatts. Thereafter, we explore the examination of other classes of IMDs that have not only higher energy requirements, but also more complicated usage and functional characteristics. The current mainstream approach in powering contemporary IMDs is to use batteries with finite lifetime [11]. Indeed, it is a known fact that the battery operational lifetime is important to all IMDs . The lifetime of the batteries in pacemakers typically lasts between 7 and 10 years, whereas the batteries in neurostimulators typically have much higher energy requirements and typically last a 3-to-5-year period before they need to be replaced [11]. Every surgical procedure required to replace a battery adds to the risk of infection and increases the cost of ownership of the device [11,12]. As a result, there is a strong impetus toward new alternative energy sources for these devices. Academics and industry professionals have been continually allocating resources toward new advances in energy storage and harvesting technologies that will extend the operational life of devices and facilitate the development of self-sustained power supplies for implantable devices [11,12]. This section provides a comprehensive discussion of the basic types of implantable devices and their energy requirements. These include, but are not limited to pacemakers, biosensors, neural stimulators, sensory prosthetics, insulin pumps, drug delivery, circulatory support systems, etc. In addition, this discourse cover current and anticipated energy harvesting and recharging approaches, as well as future trends in the energy consumption landscape for IMDs.

9.1 Pacemaker

Energy harvesters integrated with cardiac pacemakers have been thoroughly investigated because of the potential to tap into mechanical vibrations from heartbeats and convert these into electrical energy. This capability aims to facilitate maintenance-free, durable IMDs. Efforts have centered on designing robust, efficient, and adaptable piezoelectric energy harvesters that perform well under varying heart rate conditions and mechanical settings [46].

The technology outlined here involves a piezoelectric component within a cardiac pacemaker that transforms the heart's mechanical vibrations into electrical energy. This study focuses on a bistable energy harvester designed with two stable states to optimize energy capture under the heart's erratic nonlinear conditions like those caused by variable heart rates. Prior research [19,20] indicates that bistable energy harvesters can generate significantly more power than monostable ones due to the range of motions – both within a single well and amidst chaos – when subjected to regular and irregular stimuli. Animal tests using pigs and lambs were also conducted to assess the harvester's effectiveness in energy conversion from heart vibrations. Laser Doppler Vibrometry (LDV) measured vibrations at key locations such as the heart's base, apex, lung adjacent to the heart, and diaphragm. Results revealed that the heart's base is the most effective site for energy harvesting, yielding 25 μW. In contrast, the apex and lung produced less energy – 5.5 and 5.4 μW, respectively – owing to reduced vibration intensity [37,46].

Tests were conducted on the device at various heart rates, beginning from a resting state of around 70 bpm and extending to elevated levels such as 142 bpm induced by epinephrine injection. Results confirmed that the power energy harvesting (PEH) maintained excellent frequency stability, with consistent power output even amid rapidly fluctuating or irregular heartbeats, despite chaotic oscillations within [19,37,46]. Its bistable design allowed the device to effectively adapt to nonlinear excitations without compromising performance. Remarkably, the bistable harvester consistently outperformed monostable models in producing higher output power across every test scenario. Numerical analyses indicated that an 80-μm piezoelectric layer thickness was ideal, aligning the harvester's natural frequency with the heart's fundamental frequency. Furthermore, the harvester demonstrated resilience by efficiently extracting energy during both under- and overtuned conditions, with heart rates deviating from normal. Simulations project that this device can generate more than 8 μW of power under typical cardiac conditions, adequately meeting pacemaker energy requirements [20,37,46]. Additional insights on tissue dynamics were gathered from waveform measurements at the heart's base, lungs, and diaphragm. It was observed that the diaphragm and lungs function as natural dampers, mitigating heart vibration transmission. This finding emphasizes the necessity of placing the harvester near the heart's base for optimal energy conversion [37,46]. Waveform analysis showed linear increases in apex velocity and acceleration with heart rate. When standardizing these waveforms to the Kanai standard, notable differences in amplitude and shape were found across the various measurement locations.

A prototype harvester underwent testing involving an electromagnetic shaker, designed to replicate vibration patterns observed during animal trials. Additional validation was subsequently performed in this controlled setting, which is capable of replicating real-world waveforms, unlike the uncontrollable outdoor vibrations. The results demonstrated the harvester's capability to sustain the pacemaker's operation under even the harshest conditions, all while maintaining a compact size comparable to traditional implantable cardiac devices as referenced in [37,46]. This energy harvester is integrated with a prototype pacemaker, having dimensions similar to commercial ICDs. The performance integration test confirmed that this harvester produces sufficient energy to power the pacemaker over extended durations. As stated in [19,37,46], the energy harvester greatly diminishes reliance on battery replacements, thereby minimizing the frequency of invasive procedures and enhancing patient outcomes.

9.2   Implantable biosensor

Implantable biosensors are cutting-edge devices designed for continuous monitoring and diagnostics within the human body. By incorporating energy harvesters, these devices can autonomously generate power, eliminating the limitations typically imposed by battery life. They utilize methods such as piezoelectric and electromagnetic energy harvesting to derive energy from physiological actions like cardiac movements, fluid flow, or external mechanical forces [30]. Key components of an implantable biosensor include the microfluidic channel, a piezoelectric transducer, and electronics necessary for signal processing and wireless transmission. Piezoelectric materials, often made from PZT due to its high energy conversion efficiency, are commonly used in biosensors to harness energy from dynamic physiological conditions [30].

Devices relying on acoustophoresis employ standing surface acoustic waves (SSAWs) for particle or cell manipulation within microchannels, as well as for energy production. Figure 9.1(a) and (b) illustrates the resonant acoustic waves produced by a piezoelectric transducer inside the microfluidic channel, with a calculated resonance frequency exceeding 1 MHz, as derived from (9.1). The following stages in the fabrication process for such devices have been emphasized by the reference source:

$$f = \frac{C_{SAW}}{\lambda} \tag{9.1}$$

In this context, λ represents the wavelength of the standing acoustic wave and C_{SAW} indicates the velocity of the piezoelectric substrate. Microchannels are fabricated on silicon substrates using either soft lithography or photolithography methods [15,16]. The channel is then sealed through anionic bonding with a glass substrate. For final assembly, connectors and fluidic interfaces are integrated to facilitate operation and testing. Figure 9.1(c) illustrates the biosensor operating principle. A piezoelectric transducer, controlled by a programmable radio frequency (RF) signal generator, produces acoustic standing waves. These waves generate acoustic radiation forces that influence cell movement to nodes or antinodes, based on properties such as size, density, and compressibility. This mechanism allows the piezoelectric transducer to convert mechanical vibrations into electrical energy, enabling self-powered sensor operations, including the following: Cell Separation: The studies by Grenvall *et al.* [13] and Petersson *et al.* [33] demonstrate the chip's ability to isolate specific cells like circulating tumor cells (CTCs) and platelets. Cell Focusing: Acoustic forces are used for aligning cells within a designated microchannel area for subsequent analysis [14]. Energy harvesting: This involves converting mechanical vibrations into

(a)

(b)

(c)

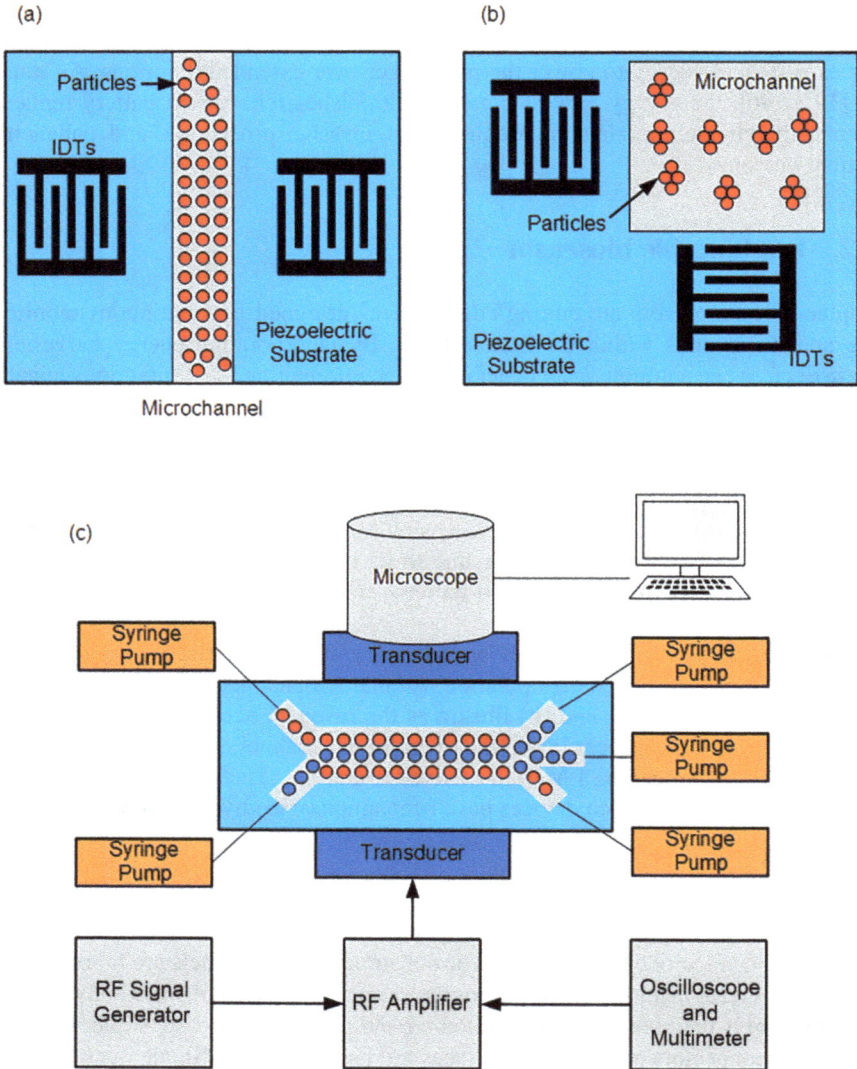

Figure 9.1   *Schematic of the standing surface acoustic wave (SSAW) devices with*
(a–b) two parallel interdigital transducer (IDTs) and two orthogonal
IDTs [30]. (c) Schematic of how biosensors combined with
piezoelectric energy harvesters can cooperate with microfluidic
devices.

electrical energy via piezoelectric materials. Figure 9.1(c) also depicts how biosensors, when combined with piezoelectric energy harvesters, function with microfluidic devices. For instance, as cells move through the channel, the SSAW device can accurately manipulate their positioning for precise analysis or separation, simultaneously harnessing energy to power the device's functions [25]. Research has explored various materials and design strategies to improve the energy density and efficiency of these systems. Examples include Cell Recovery and Viability: The use of optimal acoustic force results in minimal contamination with high cell recovery rates [36]. Label-Free Detection: Acoustic manipulation reduces the need for markers or labeling, enhancing its biological applications [2].

9.3 Implantable drug pumps integrated with energy harvesting technologies

Implantable drug delivery pumps incorporating energy harvesting technologies signify a groundbreaking advancement in the healthcare sector. These devices integrate accurate and targeted medication dispensing with self-sufficient power sources, thereby addressing the common drawbacks of traditional drug delivery methods like dose inconsistency, reliance on external energy, and frequent maintenance. With the ability to harness energy from the body, these long-lasting devices enhance patient outcomes while reducing demands on healthcare infrastructure [30]. These advanced systems rely on improved micropumps to control the flow of therapeutic agents from storage to specific locations. The most promising types are thermopneumatic, piezoelectric, electrostatic, and magnetohydrodynamic micropumps. Thermopneumatic micropumps displace fluids through volume changes created by cyclic heating and cooling [48]. In contrast, piezoelectric micropumps use an electric field to deform piezoelectric materials, producing pressure for precise fluid movement. This makes them especially appropriate for applications needing exact drug dosages [30].

Electrostatic micropumps, leveraging Coulomb's law, facilitate fluid movement by alternately attracting and repelling charged plates, offering a compact and efficient alternative for applications with space constraints [3]. Magnetohydrodynamic (MHD) micropumps employ electromagnetic fields to propel conductive fluids, thus providing an efficacious option for delivering medications. Contrarily, electroosmotic pumps utilize electric fields within ionic solutions to generate fluid flow through microchannels, ensuring exceptionally precise drug delivery [5]. In addition to pumping systems, microvalves play a crucial role in fluid flow regulation. These valves, actuated through piezoelectric, electrostatic, or thermal means, guarantee exact control over medication dosing rates. Their dynamic adjustment capabilities are especially pivotal for managing diseases requiring continuous or periodic pharmaceutical administration. Integration of energy harvesters into these systems represents significant progress. These devices convert mechanical energy, sourced from movements or heartbeats, into electrical energy powering the micropumps and valves. While thermoelectric generators (TEGs) depend on internal heat gradients, bolstering the reliability of energy supplies and thereby enabling operation under suboptimal conditions. This approach yields numerous benefits: energy harvesting prolongs device operation without regular power source replacements, reducing the need for surgical interventions; medication precision minimizes dosage errors, optimizing therapy. Conditions such as diabetes, cancer, and neurological disorders greatly benefit from the continuous delivery of personalized treatments and the alleviation of patient burdens [30]. The choice of materials for these devices carefully balances biocompatibility and functionality. For instance, piezoelectric micropumps incorporate materials resilient to physiological pressures without degradation [17]. Furthermore, advancements in microfabrication have enabled downsizing and seamless bodily integration. These devices are designed to operate autonomously with minimal patient interaction, ensuring reliability and user-friendliness. Computational modeling is instrumental in optimizing these systems. Simulations involving fluid dynamics, material properties, and energy transfer processes facilitate performance improvements and adherence to stringent medical standards. These methods also allow exploration of novel geometries, fostering further advancements in implantable drug delivery technology. In conclusion, the combination of energy harvesting and implantable drug delivery pumps represents a groundbreaking advancement in medical technology. These systems deliver patient-focused, dependable, and sustainable drug delivery solutions, addressing shortcomings of current techniques and incorporating cutting-edge advancements.

9.4   Neural stimulation implants

Neural stimulation systems are crucial in medical technology, including devices like deep brain stimulators (DBS) for Parkinson's, spinal cord stimulators for chronic pain, and vagus nerve stimulators, which is shown in Figure 9.2. They use electrical pulses to activate neural tissues, requiring more power than simpler implants such as pacemakers. DBS uses milliwatts of energy, needing lithium battery replacement every 3–5 years [11]. Spinal cord stimulators last 2–5 years, while pacemakers last 7–10 years [10,11,32]. Advances in rechargeable DBS batteries extend life to 9–15 years, requiring weekly or monthly external recharging [1,8,21]. This poses a challenge, requiring patients to adhere to recharging schedules [4,26]. Battery size is vital in design, particularly in confined areas like the cranial or spinal regions [11]. Effective neural implants need a reliable, compact power source to consistently deliver 1–20 mW, ensuring safety and seamless device functioning [11].

To fulfill these specifications, modern clinical neurostimulators predominantly employ internal power sources constituted of batteries that are recharged periodically through wireless methods. A frequently utilized approach is inductive coupling, where the alignment of an external charging coil over the implanted pulse generator facilitates transcutaneous energy transmission via electromagnetic induction, thereby enabling intermittent recharging of the implant's battery [27]. This method, fundamentally a variant of short-range wireless power transfer (WPT), is well-established; however, it necessitates precise alignment and can result in excessive heat generation if energy transfer is suboptimal. Research dedicated to enhancing inductive WPT predominantly focuses on refining coil designs and optimizing tuning to augment power transfer efficiency (PTE) while reducing energy absorption by adjacent tissues. For instance, a recent comprehensive review of WPT in IMDs highlighted that selecting the optimal frequency, along with designing coil geometry and alignment, is crucial for maximizing PTE at the designated implantation depth [9]. Advanced strategies,

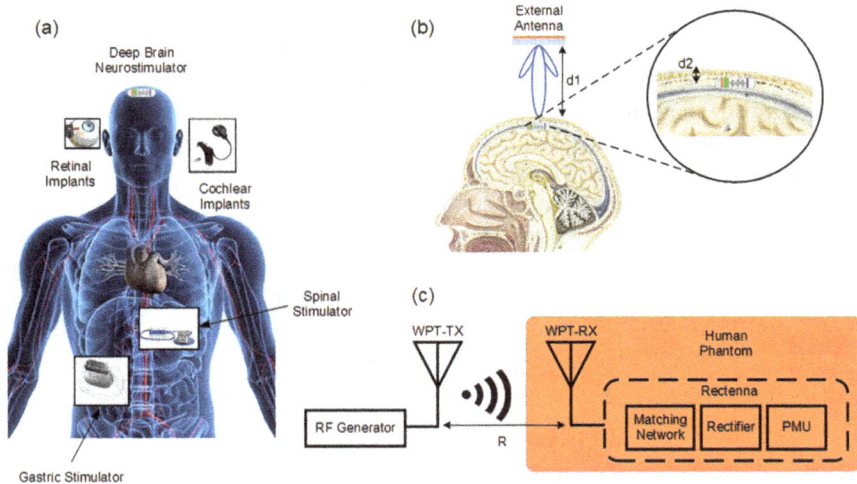

Figure 9.2 *(a) Different types of implantable stimulators [9]. (b) A wireless power transfer (WPT) process used to charge an implantable neurostimulator [9]. (c) A WPT system for stimulation application [9].*

such as employing implantable multiple-coil multiple input multiple output (MIMO) receivers and incorporating external metamaterial layers, have been proposed to mitigate misalignment losses and concentrate the magnetic field, thereby achieving enhanced charging rates for implants [9]. These technological advancements may ensure that future neurostimulators can be recharged both more rapidly and consistently, notwithstanding impeccable alignment between the internal and external coils.

Beyond rechargeable battery systems, there are persistent efforts to advance battery-free neural implants powered through energy harvesting. A noteworthy technique within this domain is ultrasonic power transfer. Ultrasonic waves possess the ability to penetrate deep biological tissues with minimal signal attenuation, and have been employed to power millimeter-scale implantable devices that do not necessitate an internal battery. For instance, "neural dust" systems – a form of miniaturized wireless sensor or stimulator – have exhibited the capability to convert incoming ultrasound into electrical energy using a piezoelectric receiver, achieving a conversion efficiency of approximately 25% in living organisms [35]. In a significant study, a neural dust device powered by ultrasound and measuring only 1 millimeter in size, successfully recorded neural activity and transmitted signals via ultrasonic backscatter, entirely energized by the received ultrasound waves [22,35]. The lack of a battery facilitates these implants to be exceptionally compact (on the millimeter scale) and durable, potentially leading to the innovation of new brain–machine interfaces and miniaturized pharmaceutical devices.

An additional innovative strategy involves the utilization of biomechanical energy derived from patient movements or organ motion to power stimulators. Triboelectric and piezoelectric energy harvesters have been integrated with or positioned on organs to convert natural body motions into electrical energy [24,35,42]. Notably, researchers have demonstrated a self-powered vagus nerve stimulation setup in rats, wherein a flexible piezoelectric harvester affixed to the stomach wall produced approximately 0.1 V pulses from peristaltic movements; this modestly harvested energy was adequate to stimulate the vagus nerve synchronously with stomach actions, achieving appetite control without the need for a battery [18,35,43]. In a similar vein, a triboelectric nanogenerator driven by respiration (a 1.2-cm device) implanted within the chest of a rat captured approximately 3.7 V from pulmonary motions, effectively energizing a pacing electrode to regulate the rat's heartbeat in real time [35,45]. These proof-of-concept implementations illustrate how, for certain low-energy neuromodulation applications, the body's intrinsic kinetic energy can be harvested to provide therapeutic electrical stimulation – essentially establishing a self-sustaining implant that draws power from its host. While these systems remain largely experimental, they propose a feasible future wherein neural implants may operate without requiring battery replacement or external recharging, by harnessing energy directly from natural physiological processes. Nonetheless, significant challenges remain, particularly in scaling the outputs of these harvesters to meet higher power requirements and ensuring the biocompatibility of these devices (e.g., employing flexible and tissue-conforming designs) [35]. The integration of advancements in rechargeable battery technology with the ongoing exploration of wireless and harvesting methodologies suggests a steady progression toward diminishing the need for surgical intervention to maintain the power supplies of neural implants.

9.5   Sensory prostheses: cochlear and retinal implants

The energy requirement from these large amounts of energy poses a major challenge to current sensory prosthesis devices, which restore partial vision in those who are blind and partial hearing in those who are deaf. These cutting-edge devices make use of large numbers of electrodes or pixels and sophisticated signal processing methods, and also directly interface to sensory nerves, which have a large power requirement. For example, cochlear implants mimic sounds by sending blocks of encoded current impulses to the auditory nerve from an internal electrode array embedded in the skull. The total energy input required for such devices is generally between 20 and 40 mW. This total energy input is necessary to provide the necessary energy to activate circuits used for the processing of sounds, in addition to those needed to provide neural stimulation [11]. Similarly, for retinal prosthetic devices, including the Argus II epiretinal implant and state-of-the-art photovoltaic subretinal arrays, an estimated 40 mW of power is required to provide the necessary energy to induce responses along neural pathways within the visual system [11]. These energy requirements cannot typically be provided by a small coin cell for an extended period of time. This is compounded by the limited spatial space available for these implants, particularly for retinal devices, within the ocular region and as a result engineers have rejected the use of large capacity batteries within these sensory prosthesis from the outset and has led to the fundamental principle of using WPT from the start.

Other medical devices harvest power externally. These include devices with an internal segment containing an electrode array and a receiver coil . The segment is embedded in a patient's head while the external segment is worn behind the ear. The external segment contains a microphone, a speech processor, and a power source (a battery). The energy and data travel across the scalp in high-frequency inductive coupling . The external unit contains a transmission coil that is magnetically aligned with a receiving coil subdermal in the implanted device . This inductive RF power transfer is the means by which all of the power needed by the implant is transferred through the thin tissue interface across several millimeters of scalp in the range of tens of milliwatts [11,44,47]. The carrier frequency for the bulk cochlear implants typically falls in the range of 2–10 MHz that allows enough power to be transferred to the order of tens of milliwatts across the thin tissue interface between the two coils. Given the close proximity and nearly perfect magnetic alignment between the two coils, the efficacy and safety of this inductive WPT has been well established for several decades . On the other hand, the downside is that the user must wear and keep in an operative condition the external unit that contains the transmitter. That condition typically includes the battery recharged or replaced daily so that the external device is ready to use when needed. Removing the external unit (e.g., during sleep or bathing) will result in the lack of functionality of the implant. As a result, there is interest within the research community for fully implantable cochlear systems. In such a system, the entire device would include the microphone, the processor, and the energy source so that no external unit would be needed when the device is in use . Since the entire energy source would be within the implant, a solution would be needed to provide the energy, typically a rechargeable battery that would be recharged transcutaneously via inductive coupling during periods such as overnight . Initial models of fully implantable cochlear systems have been constructed and tested. They have demonstrated the potential for an implant that is totally internalized and an apparatus that the patient need not remove. However, these systems tend to be bulkier to include the necessary battery and still require a period of overnight recharging via transcutaneous energy transfer (TET). Current research is focused on improving the energy efficiency of the cochlear implant electronics to reduce the size of the battery and extend the time between recharge periods [11,31].

In external power supply systems, retinal prosthetic devices are widely employed. For instance, Argus II's retinal implant has a camera and transmitter attached to eyeglasses that wirelessly transmit both visual information and power to the electrode array on the retina. Power delivery methods may be either inductive or RF telemetry; in the Argus II device, almost 6 milliwatts (mW) of power are transmitted to the implant via telemetry, where it is then converted to the needed voltage to activate 60 electrodes on the retina. Since the vision implants require continuous power during their operation, photovoltaic power has been explored as a potential wireless alternative by engineering researchers in an innovative configuration of microphotodiodes that are placed on the retinal implant that converts pulsed infrared light, emitted from video goggles, into electrical currents that stimulate retinal neurons. In this way, each pixel acts like a solar cell, with an external infrared light source, such that no implanted battery or electronic receiver is required. For example, the photovoltaic subretinal prosthesis has been developed by researchers at Stanford University, which consists of nearly 70–100 photodiode elements (each sized μm micrometers) that are each activated within the subretinal space by pulses of near-infrared light and enable visual sensations in preclinical models using only the energy from external goggles [11,35]. The photovoltaic approach eliminates the need for transcutaneous RF coils, and the number of pixels can be scaled because more current is obtained with increased light intensity, which enables more electrodes to operate. However, challenges still exist with passing sufficient light through ocular mediating, as well as safety concerns with intraocular heating, which limits the amount of available power. In 2024, WPT via inductive RF links is the most common method used in clinically approved retinal implants, with optical and ultrasound power transfer methods still under investigation.

Anticipated advancements in sensory prosthetic technology are poised to yield significant benefits through innovations at both extremes of the power spectrum. On the one side, devices are becoming more efficient at managing power consumption; on the other, there are notable advancements in wireless power delivery systems. In the realm of reducing power consumption, the introduction of state-of-the-art low-power analog and digital signal processors, the creation of refined stimulation protocols, and the application of closed-loop sensing methodologies, which strategically deactivate stimulation when unnecessary, promise to substantially cut down the energy needs for devices such as cochlear and retinal implants. Regarding power transmission, advances in WPT technologies, such as the inclusion of antenna arrays and metamaterial-enhanced transmission systems, offer the potential to increase power transfer densities, all while staying within safety regulations. For instance, simulations have shown that metamaterial surfaces can enhance energy coupling by strategically directing electromagnetic flux to the implant's receiver, which is particularly beneficial for retinal implants situated deeply within the body that typically absorb only a limited fraction of the transmitted energy [9,11]. Another promising avenue is the harvesting of ambient energy from within the sensory organs themselves. Specifically, the cochlea of the inner ear contains an electrochemical potential, known as the endocochlear potential, which is roughly 80 mV. This potential acts as a "biological battery" that has been demonstrated, in animal experiments, to generate approximately 1 nW of perpetually available electrical power, sufficient to drive an ultra-low-power sensor chip [28]. Although this power output is insufficient to directly power a cochlear implant, it underscores the possibility of extracting small trickle charges from within sensory organs. Future systems might synergistically integrate this form of energy harvesting for low-power monitoring or standby operations, complemented by intermittent wireless power bursts to accommodate more energy-intensive tasks like active stimulation processes. Taken together, cochlear and retinal implants serve as a successful example of external power supplementation, effectively eliminating the need for battery replacements, with ongoing developments focused on making the wireless energy connection more efficient, unobtrusive, and ultimately imperceptible to users, parallel to fully implanted systems.

9.6 Insulin pumps and implantable drug delivery devices

Implantable pumps for drugs such as insulin (for diabetes) require energy to drive the fluid delivery (e.g., micromotor or piezoelectric actuator) and to power the electronics (sensors, processors, etc., possibly including wireless communication). Most current insulin pump therapy uses an external pump (typically worn in a pocket or on a patch, but sometimes integrated into a continuous glucose monitor, e.g., pocket or patch continuous glucose monitor (CGM) pumps from Abbott or Dexcom) that delivers insulin through a catheter subcutaneously. These external pumps are all battery-powered: either disposable batteries that are replaced every few days, or rechargeable battery packs that avoid any surgical replacement. However, there is increasing interest in fully implantable insulin delivery systems (sometimes marketed as artificial pancreas implants, which must be completely implanted if used, and which may be implanted even when an external device is preferred). An implantable insulin pump requires either a battery or wireless power coupling, plus percutaneous ports for insulin refills (via a needle or other puncture device). This makes it a highly implantable pump. An insulin pump's need for energy depends on how frequently and in what quantities it must deliver insulin: if it is infusing, it may need tens of milliwatts to run the motor that pushes insulin out (and possibly some monitoring), but if it is in an idle monitoring mode, the energy current could be much less. Since insulin dosing is intermittent, on average the power consumption can be a relatively low amount (many pumps might average a few milliwatts or less), although the current required to run the pump can be high for short durations. Additionally, if the insulin pump includes a continuous glucose monitor and closed loop controller to make the full closed loop artificial pancreas, this adds a continuous current load for additional sensing and computation.

To date, implantable insulin pumps have been experimental or limited to specialized use (example: Medtronic MiniMed implantable insulin pump, tested in the 2000s, had a lifetime of about 1–2 years for the internal battery and required explantation and replacement of the battery). The limited lifetime and surgery to replace the battery have prevented widespread use of implantable pumps, so good power supply is an absolute requirement for implantable drug delivery. Wireless recharging would be immediate: as with neurostimulators that are recharged transcutaneously via the skin, the battery of an implantable pump could be recharged under the skin. This approach has been used in some prototypes, allowing the pump to remain in place for several years, with the patient periodically refueling both the insulin reservoir (through a tiny implanted port) and the battery, through inductive charging. But this would require the patient to periodically fully-deplete the battery by charging it, a step they would need to remember. Energy Harvesting Approaches: Investigators are working hard to find energy harvesting approaches to assist with or even completely power implanted pumps. And, exceptionally elegant, if you think about it, with respect to the task of delivering insulin, using the patient's blood sugar as an energy source is an exceptionally neat idea. A glucose fuel cell can obtain electricity by oxidizing glucose and reducing oxygen, extracting energy from the excess glucose present in the interstitial fluid or blood of diabetic patients [35,40].

In 2013, scientists from ETH Zurich presented one possible future development in the shape of a self-powered insulin release implant that was tested with murine animals. The researchers used a biofuel cell that exploited glucose to provide power. More precisely, the device turned hyperglycemia, meaning high blood glucose, into an electric current. The electric current triggered the release of insulin from genetically engineered cells on demand as needed, presumably as seen fit by reference [35]. The authors called the cell a "metabolic fuel cell" and noted that the closed loop system was able to extract enough electrical energy from excess glucose to start secreting insulin, and switch off when glucose levels returned to normal [39]. While the electric current from this fuel cell was assured to be free of enzymes and thus potentially live longer than the few months seen in enzyme-coated glucose fuel cells, which often wear out over time [35,39,40], as of this writing the preclinical evaluations of this particular fuel cell were still in progress. Nonetheless, this line of research does show some promising potential for an implantable insulin microsystem that would be able to become energetically self-sufficient by harvesting the body's chemical energy, glucose, to maintain its own therapeutic effect [39]. In parallel, other research groups have demonstrated the prolonged functionality of implanted glucose biofuel cells in small animals. These devices produce power outputs in the few micowatts [29,35]. For example, an enzyme-based biofuel cell implanted in a rat was shown to repeatedly produce more than 2 microwatts of power for periods of at least three months in vivo [29]. This number of microwatts is far greater than what a functioning insulin pump requires, but it is possible that the wattage could be used to partially recharge a battery or to run ultra-low consumption electronic components. Given the chance to extend the lifetime of their battery, the low power requirements of these systems present the possibility of constant charging. A few microwatts could also power standby operations or run a glucose sensor in a closed loop configuration, saving the battery for running the motor when delivering insulin through the pump as needed.

Beyond the examples of biofuel cells, other sources of ambient energy have been explored for use in implantable pumps. Harvesting thermal energy from the gradients of body heat is possible using TEGs that can convert temperature gradients at various locations within the body into electrical energy. At the locus where most insulin pumps are implanted in the abdomen, there is a 2–5 °C gradient in temperature between core blood and the skin's surface; a small TEG can draw tens of microwatts from these temperature differences [29,35]. This amount of energy may be too low to drive the real time operation of a pump motor, but a TEG could continuously top off an energy storage unit. Additionally, the mechanical energy from a patients breathing or movements has the potential for being harvested via a piezoelectric element attached to the pump or worn on the torso. Generally, however, the movement of the abdomen for breathing is the only instance in which movement harvest is likely to occur.

One could envision a multi-source approach: for instance, a hybrid system using a glucose fuel cell as the primary energy source with supplemental trickle-charge from a thermoelectric harvester, all stored in a rechargeable battery that can provide bursts of power for pumping. Such integration of multiple harvesting mechanisms is an active area of research aiming to ensure power reliability for implants that have variable current demands. In the near term, the most practical solution for implantable drug pumps is likely WPT combined with improved energy storage. Inductive power links, similar to those in cochlear implants, can continuously power an implantable pump or charge its battery daily, allowing the device to be implanted indefinitely without internal battery depletion. Early studies of TET for ventricular assist devices (discussed next) have shown that sufficient power can be sent through skin for even watt-level devices, so a milliwatt-level insulin pump is well within reach of current wireless power technology (given proper alignment and coil design). The remaining hurdles are ensuring patient comfort and ease-of-use in the charging process, as well as guaranteeing fail-safe operation (the device must never completely lose power in a critical moment). As energy harvesting techniques mature, future artificial pancreas implants may attain the ideal of set-and-forget operation – drawing all needed power from the body's own metabolic and thermal resources and only requiring occasional refills of the drug reservoir.

9.7   High-power implants: mechanical circulatory support devices

In the world of modern medical implants, the devices that provide mechanical circula-tory support draw significant amounts of power, and LVADs and total artificial hearts are among the most notable examples: they require several watts of power (consid-erably more than the few milliwatts required by conventional pacemakers or neural stimulators) and sustaining a continuous blood flow means they must operate almost continuously for potentially decades, so finding a viable battery small enough for safe implantation is quite a challenge. As a result, LVADs typically do not rely on internal power either, instead making use of an external power source and using a wearable lithium-ion battery pack that connects to the implanted pump via a driveline, a wire that passes through the patient's abdominal skin. While this is absolutely necessary to keep the patient alive, the driveline presents a wound that is always open, increas-ing the risk of infection, and the system also imposes an ever-present burden on the patient of managing external equipment and keeping the battery charged. It would be preferable to do away with the driveline altogether and instead power the device wire-lessly, transferring power through intact skin. The focus of WPT for LVADs is mostly on high-power wireless transfer using inductive coupling with larger coils and possi-bly resonant coupling technology. Experimental models have already demonstrated TET systems for LVADs that can transfer tens of watts through the skin with good efficiency. The general idea usually involves embedding a flat spiral coil, around 10–15 cm in diameter, just underneath the skin (usually in the patient's abdomen or chest) and having a corresponding coil on the skin surface. The power is then transferred across the gap at resonant coupling, usually in a range of the low MHz. With care-ful tuning and alignment, the power can be transferred with efficiencies of 80% or higher, even a distance of about 2 cm [9].

These have dynamic feedback control to vary the frequency/impedance to main-tain good coupling even if the patient moves and the coil alignment changes sig-nificantly. (Safety: the high power could heat up tissue and/or induce currents in the patient's body.) Therefore, operating within limits of absorbed specific absorp-tion rate (SAR) and using efficient resonant coils, 2024 IEEE review on wireless power for IMDs notes that the optimal size, shape, and distance of the coil choice and operating frequency choice have a huge impact on the power delivered (while minimizing losses and SAR) [9]. Some experimental LVAD TET systems have adap-tive alignment, e.g., arrays of small coils that can be activated sequentially or an external movable coil that moves to center over the implanted coil so that it couples most efficiently. Other new technologies are being explored to enable more efficient high-power WPT. One option is using metamaterials – artificial surfaces that can concentrate the magnetic field from the coil. These have been demonstrated to make the inductive link more efficient. If a metamaterial layer can be placed on the external coil (or in the implant pocket), more focused power can be delivered to the implant. This is achieved by more "funneling" the power, more of the flux is funneled into the receiver (more power for same input) [9]. Another option is to have multiple receive antennas in the implant (like MIMO in communications) so that if one coil in the implant is not perfectly aligned, some of the field is picked up by the other, thus stabilizing the power reception [9]. The work discussed by Essa *et al.* on implantable antenna systems could be particularly useful for implants like artificial hearts, where interruptions in power delivery are not acceptable [9].

Prospects of Energy Harvesting: It is totally inconceivable that the total notion of using only ambient energy harvesting to power an LVAD or artificial heart will be realized in the foreseeable future, given the comparability to low-power medical implants. The biomechanical or thermal energy available in the human body is only in the range of microwatts to milliwatts, which is woefully inadequate to run a blood pump which typically requires much more power. For example, using a relatively high-efficiency wide-area thermoelectric converter, one might harvest perhaps a few milliwatts of energy from the body's heat, and if a piezoelectric device harvested energy from cardiac motion, the power output would also be a milliwatt amount output [11]. Obviously this harvested power cannot directly power a blood pump that requires several milliwatts of power. However, energy harvesting could play a secondary role; for example, energy harvesters could charge increment reserve supercapacitors or batteries in the implant. If power externally could fail, the stored energy in the heart's energy storage device could maintain the pump running temporarily. A few investigators have suggested using piezoelectric elements on the heart, not to power the pump, but to provide energy to small sensors or to top off the charge of an internal battery, to maintain pump operation in the event of a power failure of the external power source. Using harvested energy in this emergency backup role would add to the overall reliability of the system. In summary, the focus for high-power implants is on wireless power transmission to the device, freeing the patient from being tethered to a skin-penetrating transcutaneous power cable. Using wirelessly powered LVADs, clinical results show that it is technologically possible to maintain pump operation for several hours using inductive coupling. The remaining challenges for wireless power reception are to ensure the long-term safety and reliability of the implanted coil (i.e., the coil will not heat up or be encapsulated by fibrotic tissue over an extended period of time) and effectively manage the external power apparatus (the batteries and the coil the patient must carry). As the high-power WPT system matures, we would expect that the future artificial hearts and assist pumps would be fully implantable, with no skin-penetrating cables protruding through the skin, significantly reducing the risk of infection and greatly improving the quality of life for the patients. Any extra energy that could be scavenged from the body (whether thermal, kinetic, etc.) would be a welcome addition for more efficiency or contingency power. However, for the foreseeable future, the main source of energy for these devices would still come from external sources of electromagnetic energy, efficiently interfaced with the implant.

9.8   Trends and future directions in implant powering

A unifying goal is to minimize or eliminate the need for battery replacement surgeries. Strategies to achieve this include developing high-capacity energy storage (improved batteries or supercapacitors) that can last the entire functional life of the device, and integrating energy harvesting so that implants become at least partially self-recharging. As noted in a recent Advanced Materials review, addressing the energy challenge is critical for the next generation of wearables and IMDs, and recent advances in materials for batteries and harvesters aim to "eliminate the need for battery replacements" by enabling self-powered operation [12]. We are already seeing commercial devices with significantly extended battery life (e.g., 15-year rechargeable neurostimulators), and at the same time, research prototypes that forgo batteries altogether (e.g., biofuel-cell-driven insulin release implants). In practice, many future implants may adopt a hybrid approach: using a smaller rechargeable battery that is continuously topped up by one or multiple harvesting modalities (mechanical, thermal, biochemical), thereby greatly extending intervals between external recharges or replacements [11]. This concept of "sustainable implants" is reinforced by the growing body of research showing that even modest energy scavenged from the body can yield significant improvements in battery longevity when intelligently managed. [11,34]

Achieving viable energy harvesting and storage in the body requires innovations in materials science. Energy harvesters (piezoelectric, triboelectric, thermoelectric, photovoltaic, etc.) must be made more efficient, flexible, and biocompatible than their traditional counterparts. For example, piezoelectric nanogenerators based on polyvinylidene fluoride (PVDF) or novel biocompatible composites can produce higher outputs under tiny deformations, suitable for implantable use [35]. Triboelectric generators are being built with nanoscale textured surfaces and soft polymers to boost charge generation from even subtle motions, while also using bioresorbable materials (like silk fibroin or poly(glycerol sebacate)) to serve short-term implant needs [35,38].

On the storage side, research into biocompatible batteries focuses on solid-state electrolytes and encapsulation techniques that prevent any toxic leakage, as well as exploring biodegradable batteries for temporary implants that dissolve once their mission is complete [23,35]. The 2021 review by Sheng *et al.* highlights the emergence of fully bioabsorbable energy devices – for instance, a transient triboelectric nanogenerator made from natural polymers (silk, cellulose, chitin, etc.) that can power a device for a few weeks and then harmlessly dissolve, avoiding the need for removal surgery [6,7,35,41].

Such developments align with the trend toward "green" implants that not only harvest renewable bodily energy but are also made of eco-friendly materials. Furthermore, advanced power management circuits are being developed (often in CMOS technology) that can operate at nanowatt levels to efficiently gather and store energy from these new harvesters [28]. All these materials and electronics advancements contribute to improving the net energy balance of the implant – either by extracting more energy from the environment or by wasting less. **Enhanced WPT:** As discussed for each category, wireless energy transfer is a cornerstone technology enabling sealed, long-term implants. The broad trend is toward higher efficiency, longer range (deeper implants), and smarter control in WPT systems. This includes exploration of frequencies beyond the common inductive (<10 MHz) links – for instance, mid-field and far-field techniques at GHz frequencies using highly miniaturized implantable antennas. While far-field RF transmission can send milliwatt-level power over longer distances, the challenge is that only a tiny fraction of radiated energy may be intercepted by a small implant antenna deep in tissue [9]. Research is ongoing into directional antennas and external beamforming to selectively target implants and improve absorption. The use of relay or repeater coils placed subcutaneously is another idea to help channel energy to deeper implants. Ultrasound-mediated transfer, as already utilized in neural dust, is another growing area – future systems may use external ultrasound transmitters aimed at implanted receivers to charge devices like leadless pacemakers or small sensors located deep in the body where RF penetration is poor. Each of these approaches must contend with safety limits (e.g., FDA limits on ultrasound intensity and on electromagnetic SAR in tissue), so innovations like adaptive power control and automatic foreign object detection (to prevent heating of unintended targets) are being incorporated. The trend toward closed-loop power control ensures that an implant receives just the power it needs and no more, adjusting transmitter output in real time – this not only saves energy but also avoids excessive fields around the implant. The inclusion of multiple coils and metamaterials in WPT, as noted earlier, represents a state-of-the-art improvement to enhance robustness and efficiency of energy coupling [9]. We anticipate that as these techniques mature, wireless charging of implants will become faster, more forgiving to alignment, and capable of powering multiple implants in the body simultaneously (through techniques like time-multiplexed or frequency-separated power delivery to different devices).

Powering implantable devices is a topic that has seen exponential growth in research attention. A recent bibliometric analysis of energy harvesting for IMDs by Fuada *et al.* documents a continuous increase in publications over the past 15 years, reflecting how critical and vibrant this field has become [11]. The analysis highlights that researchers worldwide – led by contributions from the United States, Europe (e.g., University of Bern), and China – are actively developing solutions, and that collaboration across engineering, materials science, and medicine is intensifying [11]. Key research clusters identified include work on nanogenerators (piezoelectric/triboelectric), biofuel cells, and wireless charging technologies, all aimed at improving implant power autonomy. The involvement of industry is also notable, as the market for active medical devices (wearables and implantables) is projected in the tens of billions of dollars, creating commercial impetus for better power solutions [12].

This has led to partnerships between academia and device manufacturers in developing next-generation power sources, such as implantable fuel cells or ultra-low-power chips. **Toward Self-Powered and Smart Implants:** Ultimately, the convergence of the above trends leads to the vision of self-powered smart implants that can operate indefinitely by harvesting energy, and that intelligently manage their power use. In the near future, we expect to see more implants that are energy-neutral – for example, a smart sensor that continuously monitors a physiological parameter and never needs a battery change because it scavenges energy from the body's heat or motion. For active therapy devices (stimulatory or drug-delivery), partial self-powering will extend battery life significantly, perhaps requiring recharging only on much longer intervals. The reduction of surgical battery replacements and of patient maintenance burden (no frequent recharges) would represent a major improvement in the standard of care and quality of life for patients. Furthermore, self-powered implants align with the expanding vision of the Internet of Medical Things (IoMT), where networks of body implants and wearables communicate continuously. Energy harvesting will be a key enabler for the IoMT, ensuring that a multitude of sensors and actuators can function without overwhelming maintenance needs. There are certainly challenges ahead: energy harvesters must be made even more efficient and small, energy storage must be safe and long-lived, and regulatory standards for new power modalities (like ultrasonic charging or biochemical power sources) need to be established. Nonetheless, the progress surveyed in this chapter – from successful animal demonstrations of battery-less implants to new recharge modalities for clinical devices – clearly indicates that the field is moving steadily toward the goal of "plug-free" implants. In summary, the future of powering IMDs will likely be characterized by multi-source energy solutions (combining internal harvesting and external wireless power), advanced materials that make those solutions practical inside the body, and design philosophies that emphasize minimal invasiveness and maximal longevity. The convergence of these efforts will enable the next generation of IMDs to be more autonomous, safer, and easier to maintain, thereby unlocking their full potential to improve health outcomes without the current limitations imposed by energy supply.

References

[1]   R.D. Adams. Degenerative diseases of the nervous system. *Principles of Neurology*, pages 1046–1107, 1997.

[2]   M. Antfolk, C. Magnusson, P. Augustsson, H. Lilja, and T. Laurell. Acoustofluidic, label-free separation and simultaneous concentration of rare tumor cells from white blood cells. *Analytical Chemistry*, 87(18):9322–9328, 2015.

[3]   C. Cabuz, W.R. Herb, E.I. Cabuz, and S.T. Lu. The dual diaphragm pump. In *Technical Digest. MEMS 2001. 14th IEEE International Conference on Micro Electro Mechanical Systems (Cat. No. 01CH37090)*, pages 519–522. Piscataway, NJ: IEEE, 2001.

[4]   J.R .Castle, J.H. DeVries, and B. Kovatchev. Future of automated insulin delivery systems. *Diabetes Technology & Therapeutics*, 19(S3):S–67, 2017.

[5]   C.-H. Chen and J.G. Santiago. A planar electroosmotic micropump. *Journal of Microelectromechanical Systems*, 11(6):672–683, 2002.

[6]   X. Chen, Y.J. Park, M. Kang, *et al.* CVD-grown monolayer MoS2 in bioabsorbable electronics and biosensors. *Nature Communications*, 9(1):1690, 2018.

[7]   Y.S. Choi, J. Koo, Y.J. Lee, *et al.* Biodegradable polyanhydrides as encapsulation layers for transient electronics. *Advanced Functional Materials*, 30(31):2000941, 2020.

[8]   M. Deogaonkar and C.S. Machado AG. Deep brain stimulation for movement disorders: Patient selection and technical options. *Cleveland Clinic Journal of Medicine*, 79:S19, 2012.

[9]   A. Essa, E. Almajali, S. Mahmoud, R.E. Amaya, S.S. Alja'Afreh, and M. Ikram. Wireless power transfer for implantable medical devices: Impact of implantable antennas on energy harvesting. *IEEE Open Journal of Antennas and Propagation*, 5(3):739–758, 2024.

[10] Food, Drug Administration, *et al.* Summary of safety and effectiveness data (SSED). *PD-L1 IHC 22C3 pharmDx*, 2, 2015.

[11] S. Fuada, M. Särestöniemi, and M. Katz. Analyzing the trends and global growth of energy harvesting for implantable medical devices (IMDs) research – A bibliometric approach. *International Journal of Online and Biomedical Engineering (iJOE)*, 20(3):115–135, 2024.

[12] Z. Gao, Y. Zhou, J. Zhang, *et al.* Advanced energy harvesters and energy storage for powering wearable and implantable medical devices. *Advanced Materials*, 36(42):2404492, 2024.

[13] C. Grenvall, C. Antfolk, C.Z. Bisgaard, and T. Laurell. Two-dimensional acoustic particle focusing enables sheathless chip coulter counter with planar electrode configuration. *Lab on a Chip*, 14(24):4629–4637, 2014.

[14] R. Guldiken, M.C. Jo, N.D. Gallant, U. Demirci, and J. Zhe. Sheathless size-based acoustic particle separation. *Sensors*, 12(1):905–922, 2012.

[15] S. Gupta, A. Bissoyi, and A. Bit. A review on 3D printable techniques for tissue engineering. *BioNanoScience*, 8(3):868–883, 2018.

[16] S. Gupta and A. Bit. Rapid prototyping for polymeric gels. In *Polymeric Gels*, pages 397–439. Elsevier, 2018.

[17] X. He, W. Xu, N. Lin, B.B. Uzoejinwa, and Z. Deng. Dynamics modeling and vibration analysis of a piezoelectric diaphragm applied in valveless micropump. *Journal of Sound and Vibration*, 405:133–143, 2017.

[18] R. Hinchet, H.-J. Yoon, H. Ryu, *et al.* Transcutaneous ultrasound energy harvesting using capacitive triboelectric technology. *Science*, 365(6452):491–494, 2019.

[19] M.A. Karami and D.J. Inman. Equivalent damping and frequency change for linear and nonlinear hybrid vibrational energy harvesting systems. *Journal of Sound and Vibration*, 330(23):5583–5597, 2011.

[20] M.A. Karami, P.S. Varoto, and D.J. Inman. Experimental study of the nonlinear hybrid energy harvesting system. In *Modal Analysis Topics, Volume 3: Proceedings of the 29th IMAC, A Conference on Structural Dynamics, 2011*, pages 461–478. Berlin: Springer, 2011.

[21] D.J. Lee, C.S. Lozano, R.F. Dallapiazza, and A.M. Lozano. Current and future directions of deep brain stimulation for neurological and psychiatric disorders: JNSPG 75th anniversary invited review article. *Journal of Neurosurgery*, 131(2):333–342, 2019.

[22] H. Li, C. Zhao, X. Wang, *et al.* Fully bioabsorbable capacitor as an energy storage unit for implantable medical electronics. *Advanced Science*, 6(6):1801625, 2019.

[23] S. Li, K. Shu, C. Zhao, *et al.* One-step synthesis of graphene/polypyrrole nanofiber composites as cathode material for a biocompatible zinc/polymer battery. *ACS Applied Materials & Interfaces*, 6(19):16679–16686, 2014.

[24] F.C. Liu, W.M. Liu, M.H. Zhan, Z.W. Fu and H. Li. An all solid-state rechargeable lithium-iodine thin film battery using LiI(3-hydroxypropionitrile)2 as an I– ion electrolyte. *Energy & Environmental Science*, 4:1261–1264, 2011. https://doi.org/10.1039/C0EE00528B

[25] Y. Liu, D. Hartono, and K.-M. Lim. Cell separation and transportation between two miscible fluid streams using ultrasound. *Biomicrofluidics*, 6(1), 2012.

[26] A. Machado, H.H. Fernandez, and M. Deogaonkar. Deep brain stimulation: What can patients expect from it? *Cleveland Clinic Journal of Medicine*, 79(2):113–120, 2012.

[27] A. Mailis-Gagnon, A.D. Furlan, J.A. Sandoval, and R.S. Taylor. Spinal cord stimulation for chronic pain. *Cochrane Database of Systematic Reviews*, (3), 2004, Art. No: CD003783. https://doi.org/10.1002/14651858.CD00378 3.pub2.

[28] P.P. Mercier, A.C. Lysaght, S. Bandyopadhyay, A.P. Chandrakasan, and K.M. Stankovic. Energy extraction from the biologic battery in the inner ear. *Nature Biotechnology*, 30(12):1240–1243, 2012.

[29] T. Miyake, K. Haneda, N. Nagai, *et al.* Enzymatic biofuel cells designed for direct power generation from biofluids in living organisms. *Energy & Environmental Science*, 4(12):5008–5012, 2011.

[30] K. Pal, H.-B. Kraatz, A. Khasnobish, S. Bag, I. Banerjee, and U. Kuruganti. *Bioelectronics and Medical Devices: From Materials to Devices-Fabrication, Applications and Reliability*. Woodhead Publishing, 2019.

[31] J.S. Perlmutter and J.W. Mink. Deep brain stimulation. *Annual Review of Neuroscience*, 29(1):229–257, 2006.

[32] D. Persaud-Sharma, J. William Mallet, G.D. Panjeton, et al. Neuromodulation applications for chronic pain. *Journal of Medical Devices*, 15(4):040801, 2021.

[33] F. Petersson, A. Nilsson, H. Jonsson, and T. Laurell. Particle flow switch utilizing ultrasonic particle switching in microfluidic channels. In *7th International Conference on Miniaturizing Chemical and Biochemical Analysis Systems*, pages 879–882, 2003.

[34] S.M.A. Shah, M. Zada, J. Nasir, *et al.* Ultraminiaturized triband antenna with reduced SAR for skin and deep tissue implants. *IEEE Transactions on Antennas and Propagation*, 70(9):8518–8529, 2022.

[35] H. Sheng, X. Zhang, J. Liang, *et al.* Recent advances of energy solutions for implantable bioelectronics. *Advanced Healthcare Materials*, 10(17):2100199, 2021.

[36] A.J. Smith, R.D. O'Rorke, A. Kale, *et al.* Rapid cell separation with minimal manipulation for autologous cell therapies. *Scientific Reports*, 7(1):41872, 2017.

[37] S.C. Stanton, C.C. McGehee, and B.P. Mann. Nonlinear dynamics for broadband energy harvesting: Investigation of a bistable piezoelectric inertial generator. *Physica D: Nonlinear Phenomena*, 239(10):640–653, 2010.

[38] S.-H. Sunwoo, K.-H. Ha, S. Lee, N. Lu, and D.-H. Kim. Wearable and implantable soft bioelectronics: Device designs and material strategies. *Annual Review of Chemical and Biomolecular Engineering*, 12(1):359–391, 2021.

[39] N. Taylor. Scientists use implanted glucose-run fuel cell to power insulin release. https://www.medtechdive.com/news/diabetes-implanted-glucose-run-fuel-cell/646245/, 2023. Accessed: June 2, 2025.

[40] V. Vallem, Y. Sargolzaeiaval, M. Ozturk, Y.-C. Lai, and M.D. Dickey. Energy harvesting and storage with soft and stretchable materials. *Advanced Materials*, 33(19):2004832, 2021.

[41] P. Wang, M. Hu, H. Wang, *et al.* The evolution of flexible electronics: From nature, beyond nature, and to nature. *Advanced Science*, 7(20):2001116, 2020.

[42] S. Wang, L. Lin, Y. Xie, Q. Jing, S. Niu, and Z.L. Wang. Sliding-triboelectric nanogenerators based on in-plane charge-separation mechanism. *Nano Letters*, 13(5):2226–2233, 2013.

[43] L.S.Y. Wong, S. Hossain, A. Ta, J. Edvinsson, D.H. Rivas, and H. Naas. A very low-power CMOS mixed-signal IC for implantable pacemaker applications. *IEEE Journal of Solid-State Circuits*, 39(12):2446–2456, 2004.

[44] Z. Xie, R. Avila, Y. Huang, and J.A. Rogers. Flexible and stretchable antennas for biointegrated electronics. *Advanced Materials*, 32(15):1902767, 2020.

[45] G. Yao, L. Kang, J. Li, *et al.* Effective weight control via an implanted self-powered vagus nerve stimulation device. *Nature Communications*, 9(1):5349, 2018.

[46] J. Zhang, R. Das, J. Zhao, N. Mirzai, J. Mercer, and H. Heidari. Battery-free and wireless technologies for cardiovascular implantable medical devices. *Advanced Materials Technologies*, 7(6):2101086, 2022.

[47] J. Zhao, R. Ghannam, K.O. Htet, *et al.* Self-powered implantable medical devices: Photovoltaic energy harvesting review. *Advanced Healthcare Materials*, 9(17):2000779, 2020.

[48] S.J.A.F. Zimmermann, J.A. Frank, D. Liepmann, and A.P. Pisano. A planar micropump utilizing thermopneumatic actuation and in-plane flap valves. In *17th IEEE International Conference on Micro Electro Mechanical Systems. Maastricht MEMS 2004 Technical Digest*, pages 462–465. Piscataway, NJ: IEEE, 2004.

Chapter 10
Conclusion

10.1 A conclusion on kinetic energy harvesting

Kinetic energy harvesting transforms the movement of the body into electrical energy and has shown great potential to enhance both the lifespan and capability of implantable medical devices. This method is especially beneficial for devices that must operate autonomously within the body, such as pacemakers, neurostimulators, and biosensors. Due to the inherent risks of invasive battery replacement and the difficulties of regular battery swaps, creating self-powered energy systems for implant devices is crucial. In implantable technology, body motions like breathing, heartbeats, blood circulation, and joint movements are harnessed to generate electricity through kinetic energy harvesting. This approach offers a sustainable alternative to traditional batteries, as implantable devices often function in dynamic body regions. The implementation of triboelectric nanogenerators (TENGs), piezoelectric materials, and electromagnetic induction within kinetic energy harvesting allows for energy-independent implants. Each of these technologies presents distinct advantages and limitations depending on the physiological environment. Piezoelectric materials, known for converting mechanical stress into electrical energy, are frequently used in kinetic energy devices. Positioned near continuously active organs like the heart or lungs, they can capture mechanical energy from heartbeats to power pacemakers, potentially eliminating the need for batteries. They offer reliable power output and are biocompatible, making them suitable for prolonged implantation.

The primary challenge with piezoelectric materials is optimizing power output to meet the energy demands of different devices. Researchers are looking to enhance the efficiency of these materials through novel materials and microstructuring techniques, aiming to convert energy even from low-frequency body movements. This could broaden the application of piezoelectric energy harvesters in the biomedical field, making it possible to power not only low-energy devices like sensors but also more power-intensive implants. Another method of harnessing kinetic energy is electromagnetic induction that generates electricity through the relative motion between a coil and a magnet. This technique is particularly effective in environments with rhythmic motion, such as near the cardiovascular system. Cardiovascular implants can integrate electromagnetic induction systems to capture energy from arterial pulsations or blood flow, thereby providing a continuous power supply. Nonetheless, the size and design of electromagnetic harvesters pose notable challenges, especially for implanted devices where space is limited. Engineers must strike a balance between power output and device miniaturization. Innovations in microfabrication techniques are gradually overcoming these limitations, enabling the development of compact, efficient electromagnetic generators suitable for implantation.

TENGs are a relatively recent advancement in the field of kinetic energy harvesting. These devices generate electric charges when materials with distinct electron affinities come into contact and then separate. Due to their high sensitivity, TENGs are well-suited for capturing energy from minor bodily motions, such as joint movements or slight tissue deformations. A notable benefit of TENGs is their capacity to derive power from a broad spectrum of movement frequencies and amplitudes. This flexibility makes them particularly promising for use in areas of the body with unpredictable movements, where other energy-harvesting techniques might be less practical. Despite their potential, TENG technology remains in its nascent stages, necessitating further research to enhance efficiency, durability, and integration into biocompatible, long-term implantable systems.

Although kinetic energy harvesting holds considerable promise, several challenges exist, particularly for implantable medical applications. Continuous electricity supply is essential for many implanted devices to function correctly, yet variations in body movement can hinder kinetic energy harvesters from generating stable and adequate power. Ensuring reliable power delivery is vital for patient safety and consistent device operation. Implantable devices must also be small and biocompatible to minimize discomfort and immune reactions. Therefore, kinetic energy harvesters need to use materials and structures that are safe and do not pose long-term risks to human health. Engineers are exploring flexible designs and biocompatible coatings to address these challenges.

Implanted devices are often expected to function effectively over several years. Thus, energy harvesters must be robust enough to withstand the body's mechanical environment. Investigating durable and flexible materials, alongside encapsulation strategies to protect them from bodily fluids, is critical for enhancing longevity. For kinetic energy harvesting to be viable, the device's electronics must be integrated without compromising primary functionalities. For instance, a pacemaker should convert energy into a usable form without interfering with its essential operation. Power management systems play a crucial role in this integration by regulating the inconsistent output from energy harvesters and storing it for future use.

Ongoing progress in biocompatible design, energy conversion technologies, and material science significantly impacts kinetic energy harvesting in implantable devices. Key developmental areas include integrating kinetic energy harvesting with other power-generating methods, such as photovoltaic or biochemical sources, to enhance power source reliability and flexibility. Hybrid systems can harness multiple energy sources, depending on their location within the body and environmental conditions. Innovating new piezoelectric and tribo-electric materials with increased flexibility and higher energy conversion efficiency can broaden kinetic energy harvester applications. Notably, nanomaterials offer exceptional potential for boosting power density and reducing device size. Enhancements to energy harvesting systems can be achieved by incorporating wireless data transmission and remote power transfer capabilities. Radio frequency (RF) energy harvesting, for example, can intermittently recharge the device using available external power sources, reducing reliance solely on body movement. Machine learning and predictive analytics will be integrated into power management systems to improve energy efficiency by analyzing patient activity and predicting power consumption using historical data. Wearable devices, such as smart bands or patches, can complement implants by wirelessly transmitting additional kinetic energy from body movements to the implant, addressing the energy limitations of implantable medical devices. Technologies like triboelectric, piezoelectric, and electromagnetic devices can provide sustainable alternatives to short-lived batteries that require surgical replacement. Challenges persist in integrating energy harvesters with device electronics, ensuring biocompatibility, and optimizing power output.

Advancements in smart power management, hybrid energy systems, and materials science could provide solutions to these challenges. With these developments, fully autonomous implantable devices could become a reality. From this perspective, kinetic energy harvesting technologies are anticipated to evolve, offering durable, dependable, and less invasive medical devices aimed at enhancing patient care.

10.2   A conclusion on thermal energy harvesting

Thermally capturing energy, particularly from body heat and subsequently converting it into electrical power, shows significant potential for use in biomedical and implantable devices. This method can autonomously power devices that require a continuous power source, thus avoiding the inconvenience of battery replacements. By converting body heat into usable energy, it can enhance patient safety and comfort by eliminating invasive battery replacement procedures. Notably, in organs with high blood flow, such as the liver and heart, the human body provides a somewhat stable heat supply. Thermal energy harvesting systems employ thermoelectric generators (TEGs) to convert temperature differences into electrical power, capitalizing on this consistent heat gradient. TEGs often consist of materials with high thermal conductivity and efficiency to produce electricity even from minor body temperature changes. However, several technical challenges arise when designing and integrating these systems into biomedical equipment. Maintaining a consistent energy supply amidst fluctuating body conditions is difficult, and TEGs generally offer only limited power output. Due to the body's narrow natural temperature gradient, TEGs must be highly efficient. Researchers are exploring advanced materials like silicon–germanium and bismuth telluride to enhance TEG efficiency and enable adequate energy collection from small temperature differences. Additionally, concerns about safety and biocompatibility remain paramount. TEGs and their associated materials must be compatible with body fluids and tissues, as any adverse reactions could endanger device safety. Research into biocompatible materials and coatings is essential to prevent immune responses and ensure the long-term safe operation of TEGs. Efforts to develop encapsulation techniques are also underway to protect TEG components from degradation due to prolonged exposure to bodily fluids. Heat dissipation is another critical factor. Since TEGs use body heat for power generation, avoiding excessive cooling of surrounding tissues is essential to prevent interference with normal bodily functions. Effective thermal management strategies, including strategic device placement and design structures that balance heat absorption and dissipation, are needed to sustain energy production while preventing localized cooling.

Despite these challenges, thermal energy harvesting holds significant potential for implantable and biological devices. For instance, thermal energy harvesters are capable of powering low-energy sensors such as glucose, pH, or temperature sensors used in continuous health monitoring, thus eliminating the need for frequent battery replacements. Continuous monitoring is highly beneficial for patients with chronic conditions, as it allows healthcare professionals to collect real-time data, enhancing the management and treatment of these diseases. Additionally, drug delivery implants that release medication in response to specific physiological triggers show promise with thermal energy harvesting. For these devices to be effective and responsive, a reliable and long-term power source is vital. Thermal energy harvesting can provide a self-sustaining power supply, enabling drug delivery implants to function independently by adjusting medication release based on real-time biosensor data. As research advances, hybrid energy systems combining thermal energy with kinetic or photovoltaic energy are emerging as better options. These hybrid systems can offer a more reliable power supply by blending different energy sources, particularly when body heat alone is insufficient. This combination enhances both device longevity and overall stability, particularly in the face of temperature fluctuations. Continuous advancements in device design, materials, and integration with other energy-harvesting techniques are crucial to fully realizing thermal energy harvesting's potential. Ongoing research on high-efficiency thermoelectric materials, such as those utilizing nanostructures or advanced alloys, aims to enhance power output by optimizing microscale thermal and electrical conductivity. By significantly boosting energy extraction from body heat, these materials could support a broader range of implantable devices with higher power demands. Furthermore, integrating power management systems is critical for addressing fluctuations in body heat. Advanced power management circuits capable of storing excess energy generated by TEGs allow devices to function even when energy production decreases. Such systems, providing a stable power source independent of environmental conditions, would be particularly beneficial for devices in areas with significant temperature variations.

The completion of clinical evaluations in the medical field and for regulatory standards has been achieved for thermal energy-harvesting devices. These devices require extensive testing to guarantee their safety, effectiveness, and long-lasting presence in a biological environment. Regulatory pathways must address specific challenges posed by these devices, particularly their potential impact on the body's temperature regulation and the long-term safety of thermoelectric materials. Close collaboration with regulatory authorities, healthcare professionals, and researchers is crucial to establish standards ensuring that these technologies are utilized safely and effectively. Thermal energy harvesting is poised to transform the biomedical implant sector by providing a continuous, sustainable power source that eliminates the need for invasive battery replacements. By converting body heat into electrical energy, it can power a variety of low-energy devices, from drug delivery systems to sensors, thus improving patient outcomes and life quality. Despite challenges like biocompatibility concerns, power output limitations, and effective heat management, advancements in materials science, device design, and hybrid systems offer promising solutions. With further development and validation, this groundbreaking approach could revolutionize the management and monitoring of chronic conditions, leading to a new generation of energy-autonomous biomedical devices. As this field grows, it brings us closer to the vision of fully self-sustaining implantable devices that enhance patient care by delivering enduring, minimally invasive treatments.

10.3   A conclusion on bioenergy harvesting

The newly developed powering method is an implantable device based on bioenergy harvesting, tapping into the body's intrinsic biochemical pathways for generating electrical energy. This approach to bioenergy harvesting offers fresh outlooks for crafting durable, self-sustaining devices that obviate frequent battery changes or external charging by utilizing biological resources to yield usable power. This strategy is particularly beneficial for long-term implants needing continuous power, like pacemakers, glucose monitors, and drug delivery systems. Bioenergy harvesting taps directly into the body's chemical resources to power low-power biomedical devices. Unlike conventional batteries with finite lifespans that often necessitate invasive surgeries for replacement, bioenergy devices offer a sustainable internal energy source. This steady power generation allows devices to function longer, addressing chronic disease management needs and reducing repeat procedure risks. Additionally, bioenergy harvesting systems can be seamlessly integrated into the body's physiological environment. For example, glucose-based biofuel cells leverage the body's glucose and oxygen to produce electrical energy through electrochemical processes. This reaction is biocompatible, akin to natural metabolic activities, minimizing adverse or immune responses. Utilizing these biomimetic strategies, bioenergy harvesting ensures a sustained and compatible power supply tailored to medical implant demands. Nonetheless, numerous technical challenges must be conquered to ensure reliable and safe bioenergy harvesting operations. Maximizing power output remains a significant challenge since bioenergy harvesters typically produce small electricity amounts, making them suitable for devices with minimal power needs. Research into advancements in materials science, like enzyme or nanomaterial-enhanced biofuel cells, aims to enhance energy conversion efficiency to ensure even minimal biological energy meets the operational demands of implants. Another key challenge is maintaining consistent performance over time. Because bioenergy harvesters rely on biological processes within the body, variables like fluctuating glucose levels or variable oxygen supplies can impact power output. Researchers are developing power management technologies to control energy allocation and storage, ensuring device functionality even during periods of biofuel scarcity.

Bioenergy harvesting in implants necessitates the stability and biocompatibility of devices. Biofuel cell components must endure the body's harsh chemical conditions without degradation and should not trigger immune responses. To ensure long-lasting stability and reliability, the development of durable, biocompatible materials is vital. Researchers are also exploring protective coatings and encapsulation techniques to shield biofuel cells from enzymatic breakdown and the influence of bodily fluids. The potential applications of bioenergy harvesting in a variety of implantable medical devices are promising. For example, biofuel cells powered by glucose could enable continuous glucose monitors to operate autonomously by using the very molecule they measure as their energy source. This approach could drastically reduce the bulk of devices by eliminating the need for large batteries, thus enhancing patient comfort and reducing invasiveness. Moreover, bioenergy harvesting presents a reliable alternative to batteries in cardiac pacemakers and other heart-regulating implants. Batteries have limited lifespans and necessitate surgical replacement. A glucose-fueled biofuel cell implanted in a pacemaker could tap into the body's glucose reserves to generate the small but steady power needed to maintain cardiac rhythm. This would heighten the pacemaker's independence, reduce the need for replacements, and improve patients' quality of life. Likewise, bioenergy harvesting could enhance controlled drug delivery in implantable systems by harnessing the body's metabolic resources. These systems use real-time biosensor data for accurate energy management to adjust medication dosages and schedules. With bioenergy harvesting, these devices can operate for extended periods without external power, offering more personalized, adaptive treatment options for conditions like diabetes, cancer, and pain management.

The advancement of new material studies, the enhanced efficiency of biofuel cells, and the integration of hybrid energy systems are all motivated by the potential of bioenergy harvesting for implantable medical devices. This crucial area of enzyme-based biofuel cells has seen ongoing enhancements in enzyme stability and durability. While enzymes in these cells facilitate metabolic processes, they are gradually degraded by the host body. To extend the lifespan of biofuel cells for long-term implantation, efforts are underway to improve enzyme stabilization and immobilization techniques. Additionally, nanotechnology shows significant promise for boosting bioenergy harvesting efficiency. By incorporating nanomaterials like graphene or carbon nanotubes, biofuel cells can achieve greater surface areas, enhancing energy conversion efficiency and electron transfer rates. In the energy-limited environment of the human body, biofuel cells with nanomaterial enhancements could power a broader range of implantable devices, including biosensors and neurological stimulators. Hybrid energy systems, which combine bioenergy harvesting with other energy capture methods like thermal or kinetic harvesting, offer comprehensive power solutions for biomedical devices. By harnessing multiple energy sources, hybrid systems can compensate for fluctuations in biofuel supply, providing a more reliable and stable power source. This strategy is particularly well-suited for devices operating in various physiological conditions, where the availability of a single energy source might vary with patient activity or metabolic state.

Initially, these biomedical implants signify a groundbreaking application of bioenergy harvesting by providing a self-sustaining, biocompatible, and renewable energy source that aligns closely with the body's biochemical processes. By leveraging glucose and other biological mechanisms for energy acquisition, this method of bioenergy harvesting minimizes the need for frequent invasive battery replacements and enhances patient comfort, while ensuring the continuous operation of devices through less intrusive methods. Thanks to ongoing progress in material science, nanotechnology, and the integration of hybrid systems, advancing more effective, durable, and reliable bioenergy solutions is becoming feasible, despite challenges related to power output, stability, and biocompatibility. As these innovations progress, bioenergy harvesting is poised to become a central element in the next generation of energy-autonomous medical devices, allowing for tailored treatment that minimally disrupts patients' everyday lives. This approach foresees a future where implantable devices can sustainably operate within the body, providing patients and healthcare providers with devices that are as robust and adaptive as the biological systems they support.

10.4 A conclusion on RF energy harvesting and wireless power transfer

10.4.1 A conclusion on RF energy harvesting

RF energy harvesting, which transforms ambient radio waves into usable electrical power, is gaining attention as an innovative solution for powering implantable and biomedical devices. This method delivers a continuous, noninvasive power source by harnessing energy from nearby RF sources, such as cellular networks, Wi-Fi routers, and dedicated RF transmitters. This feature is particularly beneficial for implantable devices operating autonomously within the body, as it reduces the dependence on conventional batteries that require regular replacement through invasive procedures. In specific scenarios, RF energy harvesting permits implantable devices to draw power from ambient RFs, eliminating the need for bulky batteries. By tapping into broadly available signals, RF energy supports the constant operation of devices, which is crucial for low-power sensors and communication modules. This steady power generation is especially advantageous for small, minimally invasive devices like drug delivery implants and biosensors that often demand only a modest and reliable energy supply to function effectively..

One significant benefit is the capability for wireless data transfer made possible by RF energy harvesting. Numerous biomedical implants use wireless communication to send patient health data to smartphones or external monitors, facilitating real-time remote monitoring. The adoption of RF energy harvesting enables these such devices to utilize a single power source for both operation and data transfer, eliminating the need for wires and frequent recharging while also reducing the device size for greater patient comfort. Despite its promise, RF energy harvesting in implantable biomedical devices faces multiple challenges. The primary issues involve power consistency and availability. The extractable energy from RF sources is often restricted and can fluctuate due to factors like signal strength and proximity to RF transmitters. Power inconsistencies can arise when a patient moves out of

RF range or when body tissues influence signal strength. To address these issues, advanced power management systems are crucial for optimizing energy capture and storage, ensuring device functionality even if RF energy temporarily decreases. Signal weakening and interference from tissues pose additional complications. As RF waves traverse body tissues, absorption and scattering cause them to diminish in intensity, thereby reducing the energy reaching the implant. Furthermore, different tissues, such as skin, muscle, and fat, each absorb RF energy differently and complicate power efficiency. Researchers are developing specialized antenna designs and resonance-matching techniques to enhance RF energy capture within the body. These innovations aim to increase the amount of RF energy reaching implants, even under challenging conditions.

In addition to these challenges, RF energy harvesting offers the potential to power an array of medical devices. For example, biosensors involved in continuous health monitoring could utilize RF energy harvesting to supply constant power, enabling the real-time detection of vital indicators like blood oxygen saturation, heart rate, and glucose levels. Utilizing RF radiation to power these sensors relieves patients from the hassle of frequent battery replacement, while providing medical professionals with continuous data for informed treatment decisions. Applications such as drug delivery systems also show significant potential with RF energy harvesting. Accurate energy management is often required by implanted drug delivery systems for controlled drug release. As they can operate independently with RF energy harvesting, these systems eliminate the need for an internal battery, allowing drug release to adjust in real time based on physiological data. This capability is especially advantageous for treating chronic diseases, enabling personalized, consistent treatment that requires less maintenance. Additionally, RF energy harvesting may facilitate neurostimulation devices that address conditions like Parkinson's disease, epilepsy, and chronic pain. These devices deliver controlled electrical impulses to specific brain regions and typically need steady, small amounts of power. The financial and physical burdens on patients could be reduced if neurostimulators could harness RF energy autonomously, eliminating the need for frequent battery changes or recharging. The advancement of RF energy harvesting for biomedical purposes is poised for significant growth due to innovations in engineering and materials. A promising research area is the development of compact, highly efficient antennas designed specifically for implants. These antennas can be tuned to harvest energy from specific frequency ranges, such as cellular networks or Wi-Fi, to optimize RF energy harvesting. With optimal antenna designs, implantable devices can more effectively capture and convert RF energy even from distant sources or in low-signal environments. Furthermore, hybrid energy systems, which integrate RF energy harvesting with other energy sources like thermal or kinetic energy, address various power needs. By using RF energy when available and shifting to alternative power sources as needed, hybrid systems enhance device reliability throughout its operational lifespan by maintaining functionality across different conditions without reliance on a single energy source. There is also an urgent need for advanced power management circuits that can store and regulate energy acquired from RF waves. These circuits aim to easily transfer electricity from an antenna to the device's active components while minimizing the impact of power-intensive functions such as wireless data transmission. By storing excess RF energy during strong signals, these systems ensure a steady power reserve, allowing devices to function briefly during potential interruptions in RF availability.

10.4.2 A conclusion on wireless power transfer enhancement

The utilization of wireless power transfer (WPT) for RF energy harvesting is paving the way for innovative methods to power implantable and biological devices. As an emerging approach in wireless power delivery, WPT uses electromagnetic fields to transfer energy over short distances. This not only facilitates RF energy collection but also ensures a more stable and concentrated power source for medical implants. By integrating WPT with RF energy harvesting, it becomes possible to continuously and reliably power implanted devices, reducing the invasiveness of conventional battery replacements. Combining WPT with RF energy harvesting offers numerous benefits for implant powering. WPT systems are engineered to efficiently transfer energy directly to implants via capacitive, resonant, or inductive coupling. This is particularly advantageous for devices implanted at shallow depths, like subcutaneous biosensors or cochlear implants, where energy can be supplied directly through the skin from external power sources. WPT allows for steady power supply without relying on battery storage limitations, meeting the energy demands of devices as they arise. Medical professionals can precisely control and adjust power delivery, catering to devices with variable power needs, such as drug delivery systems or neurostimulators, to ensure optimal device performance. This approach is vital in medical applications, promoting personalized patient care and enabling remote monitoring and management of device functions. By combining WPT with RF energy harvesting, implantable devices gain flexibility and reliability, forming a dynamic system while enhancing overall dependability. RF energy harvesting involves collecting ambient energy from diverse RF sources, and WPT can provide a high-efficiency power source as needed. If RF energy falls short or is unpredictable, WPT can become the primary or backup source. This hybrid strategy lessens dependence on a single energy source, extending device life and ensuring continuous power supply. It allows for the inclusion of a small implantable battery or supercapacitor, recharged via WPT, to serve as a power reservoir when RF energy is scarce. This arrangement enables devices to seamlessly switch between active and standby modes, depending on power source availability, which is beneficial for implants in fluctuating RF environments where patients transition between areas with varying RF signals. Despite the many advantages of integrating WPT with RF energy harvesting, there are some challenges. WPT must emphasize power efficiency and safety to avoid interference with other medical devices and prevent overheating. Researchers are developing WPT systems with precise targeting and low-frequency strategies to focus energy on specific implants while minimizing exposure to surrounding tissues. Optimizing WPT and RF energy capture requires careful antenna design, ensuring they are tuned to match the external power source's frequency for effective energy transfer. RF antennas must efficiently capture ambient energy within the body. Advances in resonance-matching and antenna miniaturization are helping to overcome these challenges and enable effective energy transfer in the body's complex environment. Important considerations also include biocompatibility and thermal regulation. The integration of WPT and RF systems requires durable coatings and materials to prevent immune reactions and avoid overheating. Biocompatible covers are being developed to dissipate heat and insulate implant components, maintaining safe temperatures and protecting surrounding tissues from thermal damage.

Applications of the integrated WPT and RF energy harvesting in biomedical devices include the following:

- **Cardiac devices and pacemakers:** Heart-related devices such as pacemakers and other implantable gadgets require a constant and dependable power supply. Typically, these devices are powered by a hybrid energy solution in regular operation. WPT can be occasionally utilized for recharging, which decreases the frequency of battery replacements and related surgeries, whereas RF harvesting can consistently provide low-power energy.
- **Drug delivery devices:** By combining WPT with RF energy harvesting, drug delivery implants can utilize a stable power supply for precise and controlled drug release. RF energy harvesting maintains the device's basic functions, while WPT provides power on demand for specific drug dosages.
- **Health monitoring systems and biosensors:** This hybrid power approach is highly advantageous for continuous health monitoring systems. WPT adds additional power when significant data transmission is required, while RF energy harvesting supports standard sensor tasks such as monitoring glucose or oxygen levels. This setup enables real-time monitoring without the need for regular power replenishment.
- **Neurostimulation devices:** WPT provides reliable, controlled energy for devices that necessitate occasional bursts of power, like neurostimulators used for epilepsy treatment or pain management. The device can remain functional between WPT sessions by utilizing RF energy harvesting to capture surrounding energy for low-power tasks. To achieve the full potential of WPT and RF hybrid systems, improvements in miniaturization, efficiency, and biocompatibility are crucial. Future research will likely focus on developing versatile antennas capable of supporting both WPT and RF harvesting, allowing implants to seamlessly alternate between power sources. Additionally, dynamic power management systems that can adjust energy intake and distribution according to available sources will be vital for maximizing efficiency and device lifespan.

Another promising area is the application of AI and machine learning algorithms to enhance energy management. AI-driven systems can anticipate energy needs by analyzing consumption patterns and environmental conditions, enabling decisions on when to activate WPT or switch to RF harvesting for optimal efficiency and continuous operation.

RF energy harvesting represents an innovative solution for powering implantable biomedical devices by offering a sustainable alternative to traditional batteries. The effectiveness of RF harvesting lies in its ability to convert ambient radio signals into a usable energy source, allowing continuous device operation and potentially reducing the need for regular battery replacement surgeries. This method is particularly effective for low-power implantable devices, like drug delivery systems and biosensors, that demand only a minimal and consistent power supply to support data transmission and other operations. By integrating WPT with RF energy harvesting, new advancements in biomedical implant power sources are achievable. This hybrid approach is ideal for applications requiring periodic bursts of energy, as it harnesses environmental RF energy for ongoing low-power demands while utilizing WPT for targeted, reliable power supply. This combination reduces reliance on traditional batteries and contributes to a sustainable, long-term solution for implantable devices, eliminating invasive surgical procedures. Challenges such as heat regulation, maximizing power efficiency, and ensuring biocompatibility remain, but ongoing research and technological improvements are addressing these issues. The future of implantable devices lies in fully autonomous, energy-resilient systems that seamlessly unite WPT and RF harvesting. As this field advances, it will usher in a new era of medical implants, where devices operate continuously, adapt to changing conditions, and deliver personalized care with minimal patient intervention, transforming medical technology.

10.5 A conclusion on photovoltaic energy harvesting

The development of the implantable photovoltaic energy harvesting system is completed with device fabrication, on-chip power management circuits, and protective encasings. To shield the PV cell from subcutaneous fluids, polymer encapsulation and a hermetic package are applied. Implantable PV cell fabrication frequently involves the use of monocrystalline silicon, alongside CMOS technology in device creation. Optimizing the use of a junction diode is crucial for boosting photovoltaic cell efficiency. CMOS technology provides benefits such as reducing the system size and allowing the integration of photovoltaic cells with CMOS circuits. The focus is on the packaging and encapsulation. It is essential to note that while a solid hermetic structure offers excellent protection for the devices, polymer encapsulation is more adaptable and comfortable. Silicone is among the top candidates owing to its excellent optical transparency, flexibility, and compatibility with optical semiconductor devices in advanced manufacturing. Factors like the light source and optical skin loss also impact the performance of a PV cell. If the issue of heating is addressed, NIR light has proven to be an effective input power source for implanted PV cells. Human tissue characteristics vary based on ethnicity, geographic location, age, and even different parts of the same individual. Additionally, evidence suggests that PV cells implanted in the hypodermis deliver more stable energy compared to those in the dermis, which can harness more dynamic power. Implantable PV cells outshine other power harvesting techniques by providing sufficient power in a compact form. The PV cell's circuitry is simpler than those of other AC power sources, and it is compatible with embedding PV cells on the same chip. Nonetheless, implantable PV cells face challenges, such as limited light feasibility and intensity, high costs, inflexibility, the necessity for more research into long-term performance post-implantation, and the need for deeper implantation within tissue surfaces.

10.6   A conclusion on hybrid energy harvesting

This chapter examines hybrid energy harvesting techniques aimed at delivering efficient, reliable, and sustainable power supplies for biomedical and implantable devices. The focus is on hybrid systems that integrate the strengths of various single-source energy harvesting methods to address limitations and provide robust solutions suitable for a range of environments, from low-energy to highly dynamic settings, particularly within and around the human body. Initially, the study explores different hybrid energy harvesting systems, materials, and structural designs, emphasizing the need to select suitable materials and optimize structures to boost efficiency and energy output. By combining materials with complementary properties, hybrid systems can capture energy from multiple sources, thereby enhancing the total energy harvested while maintaining compact, implantable, and biocompatible forms. The discussion then shifts to piezoelectric–electromagnetic (PE–EM) hybrids that marry electromagnetic induction from the movement between a coil and a magnet with the piezoelectric effect of mechanical strain. These systems are ideally suited for powering low-energy medical devices through regular body motions, such as limb movement or heartbeat, providing a consistent power output adaptable to varying motion frequencies and amplitudes. Additionally, the chapter explores other hybrid energy harvesters that merge traditional mechanical methods with less conventional sources like thermal, biochemical, and solar energy. By harnessing ambient sources such as light, chemical interactions, and body heat, these devices broaden energy collection possibilities, thus enabling specific applications where devices can sustain themselves using energy from diverse physiological and environmental sources. In conclusion, hybrid energy harvesting technologies offer reliable, adaptable, and efficient energy-generation options, marking a significant advancement in powering biomedical devices. These hybrid systems enhance energy output and mitigate the shortcomings of individual harvesting methods by amalgamating various mechanisms. They pave the way for sustainable, self-powering medical implants and wearables that minimize the reliance on traditional batteries, boosting patient comfort and safety. Future research in biocompatibility, materials science, and miniaturization will further refine these hybrid systems, unlocking innovative medical solutions that seamlessly integrate with the human body's dynamic energy landscape.

Index

www.ingramcontent.com/pod-product-compliance
Lightning Source LLC
Chambersburg PA
CBHW050517190326
41458CB00005B/1571